MAPS AND MEANINGS

MAPS AND MEANINGS
URBAN CARTOGRAPHY AND URBAN DESIGN

JULIE NICHOLS

ACADEMICA PRESS
BETHESDA - DUBLIN - PALO ALTO

Library of Congress Cataloguing-in-Publication Data

Nichols, Julie.
Maps and meaning : urban cartography and urban design / Julie Nichols.
pages cm
Includes bibliographical references and index.
ISBN 978-1-936320-65-3
1. Cartography--Philosophy. 2. Cities and towns--Maps. 3. City planning. I. Title.
GA102.3.N53 2014
526--dc23
2013049245

Academica Press, LLC
Box 60728
Cambridge Station
Palo Alto, CA. 94306

Website: www.academicapress.com

to order: 650-329-0685

For my lovely family: Amelie, Xavier, Sebastian and Mark.

CONTENTS

LIST OF FIGURES

FOREWORD

In the second half of the tenth century, a Muslim merchant from Baghdad known as Ibn Ḥawqal (d. c. 988) took it upon himself to map the entire earth. To do so he had to circumnavigate the earth, lands and seas, in order to draw and record first hand, everything he could see and experience. He began his ambitious expedition in May 943 and it took him nearly thirty-five years to complete his mission. Around 977, three years after his return, he completed his famous book *Ṣurat al-Arḍ* (*The Image of the Earth*), which contained a large map of the entire earth as well as detailed regional maps of all known regions between China in the east and Andalucía in the west. In addition, he provided extensive textual descriptions of the geographies and cultures of the lands and peoples he mapped. Ibn Ḥawqal did not start from scratch, of course, he had access to previous maps and geographical information found in Arabic, Indian, and Greek sources, yet his new and authoritative maps looked more like imaginative children's drawings than proper geographical documentations. For him and for the people among whom his book was circulating, however, these maps represented geographical reality, a true image of the earth. His atlas represented the most "accurate" and detailed mapping available then, one that was verified by an eyewitness experience on the ground. The maps graphed a mode of seeing more than what really existed out there; nevertheless, they were still real and accurate. After thirty-five years of tireless travel and diligent observation, the cartographer's maps and words remained closely tied to his personal vision and experiences, to his mode of seeing and representing, and could not stand alone as objective representations of independent geographical reality.

Ibn Ḥawqal's experience reveals a predicament inherent in the act of mapping. As a representational act, mapping always generates a duality of the real and its image. This duality presents a dilemma: which comes first, the geography or the map? Today, it is taken for granted that the territory comes before the map, and the result is that geographic reality has become objective and spatially stable. But this has not always and everywhere been the case. In Ibn Ḥawqal's case, no-one then knew what the territory looked like, thus his maps preceded the territory. In fact it created the territory. His image of the entire earth realised the territory: it made it real, visible and tangible. In this case, when the map precedes the territory, the geographic reality can never be spatially stable, never something external and given for us to see. For this reason, maps were originally conceived as a means of finding and founding the world. Until modern times, when modern technologies have facilitated the projection of stable, non-negotiable images of the earth, mapping was largely an individually creative act. Pre-modern cartographers were able to creatively project their own images of the world, and these projections were not mimetic or merely descriptive. They were creative projections, tracing a potentiality rather than a fixed reality, and unfolding in a multitude of forms according to different engagements and visualizations.

To *map* is to take the measure of the world. Taking measure can be personal, as in the case of Ibn Ḥawqal, and can be impersonal, as in modern surveys, yet in both cases they involve, on the one hand, selecting, translating and differentiating, and on the other, visualising, conceptualising, recording and representing. Thus mapping is first and foremost a projection of order, of a particular order, of a particular logic and way of seeing. Whether subjective or objective, conceptual or graphical, mapping configures space, translating it into a familiar and recognisable place wherein geographic locations become meaningfully related. Mapping is not confined to the registered and the drawn; it can be spiritual, political or moral, and it can include the remembered, the imagined, the anticipated and the desired.

This book tells the story of mapping in its modern and premodern manifestations. It takes the reader on a journey to explore the meanings and utility of mapping in different cultural and professional contexts. Focussing on Southeast Asian cartography, it provides insights into modes of seeing and representing, totally alien to modern western conventions. It shows how the developments of the European Enlightenment and rise of modern science have radically changed the nature of mapping, and how the new mapping approaches and techniques have become instrumental in the modern practices of town planning and urban design. It also shows how the quality of urban environments could be seen as a reflection of mapping conceptions and practices. It is a rich and intellectually stimulating study that provides new insights into the history and theory of cartography and their role in representing, realising, and giving meanings to human presence and human settlements.

Professor Samer Akkach
Adelaide

PREFACE

In today's practices of urban design, the map acts as a documentary and design tool as well as a legal document. Its usefulness hinges on its perceived truthfulness and objectivity in the representation of reality. Yet this has not always and everywhere been the case. There was a time in Western and non-Western societies where the nature of the map and the acts of mapping were very different. This study traces this difference in an attempt to understand the process of change and its impact on the nature and quality of human settlements.

To do this, the study explores points of intersection between urban design and urban history. Focussing on Southeast Asia, it examines the transition from pre-modern to modern modes of mapping enabled through the mediation of Western intervention. The aim is to comparatively trace the map's historical evolution in intertwining Western and non-Western contexts. Using archival materials, the study brings together Southeast Asian urban history, history of urban cartography, and urban design theories. It shows how different forms of mappings reveal culturally specific ways of seeing and understanding the world. Pre-modern maps typically prioritised sacred and profane space and the proliferation of religious knowledge over the need to satiate any geographical enquiries. As technological developments in Europe brought about new forms of cartography, Western ideas about space, previously dominated by socio-religious beliefs, were openly challenged by science and exploration. The Enlightenment period's embrace of reasoned knowledge and rational thought filtered into mapping practices, which was eventually embraced globally to the demise of sacred space. Yet the past survived in urban history, and between the retrospective view of urban history and the projective view of urban design a new schism emerged.

By examining the role of the map at a conjunction of urban history and urban design, the study attempts to show how the Enlightenment's rational mapping proliferated into the non-Western world, how the production of urban space shifted from a socio-culturally motivated style to a highly theorised framework, how the concept of the modern city was borne alongside the emergence of modern urban planning, how the emergence of modern thinking about the city corresponded with new ways of designing, and how theorists reacted to the modernist urban design rationalism which was anchored in the authority of scientific mapping. Through this path of enquiry the study strives to uncover some of the lost meanings and functions of the map, and to examine new approaches to dealing with the loss of quality and identity in today's urban environments.

ACKNOWLEDGMENTS

I am grateful to a vast number of people who have contributed to the final realization of this big project. To my mentor, Professor Samer Akkach, to whom I am indebted for so many reasons, some of which include his endurance, endless encouragement, acceptance of the many hurdles which presented themselves, and most significantly, for his continued friendship seeing the thesis evolve into a book publication. I must acknowledge Samer's commitment to refine the original thesis and yet accept any flaws that remain in this final version to be solely my responsibility. My gratitude extends to Dr Peter Scriver for his support, suggestions throughout the process and a positive outlook always at crucial times. To Emeritus Professor Terry Williamson, thanks for his humor and remarks to ensure my determination to finish. To Dr Alpana Sivam, my colleague and friend, from whom I received many valuable anecdotes and insight into scholarship. My warmest wishes go to colleagues, Dr Christine Garnaut and Peter Walker for their interest, encouragement and discussions on the topic. My sincerest thanks go to my editors of various versions, Sally Shipway, Norman Prentice, Jeff Catchlove, Myriam Sampson and Peggy Buckler for their instructive corrections and perseverance with the body of work with which they were presented. I am also very grateful to Associate Professor Veronica Soebarto and Professor Paul Babie at the University of Adelaide, Faculty of Professions, for their efforts in securing funding for copyright permissions and my professional editors. Thank you also to the professional editors for their invaluable insight and comprehensive efforts. To my research assistant Vlad Doudakliev for all of his hard work in obtaining copyrights, in addition to Catherine Daniel and her team. To all of my dear friends for their enduring support and with special thanks to Vasa Raven, Heleen Terwel, Kate Heap, Ann Laffan, Andrea Fillmore, Kong Yen, Jen Robinson, Caroline Lange, Bernie Auricht, Kerelle Paim-Bruges, Sue Drewnicki, Dougald Prentice and Paquita Kennett for listening as well as everything else. To all of my family who have assisted in every possible way from babysitting to cooking and sympathizing, I feel so very lucky. To my amazing mother, Jan, my heartfelt thanks for her love, support and belief in my abilities, to the extent that words fail me. To Peter, Belinda and Naomi, Carol, Peggy and Ted, Meg, Cathy and Matt,

Sue and Ian, Dave and Belinda, Tim, Lori and Marc and all their children, it is so wonderful to have you as my family. Finally, a huge thank you, to my gorgeous children, Amelie, Xavier and Sebastian for their cheekiness, love, cuddles always and who make my world complete, and finally to Mark for his love, patience with this project, and natural aptitude for "holding the fort!"

INTRODUCTION

An early career posting in Kuala Lumpur, Malaysia, fostered my enthusiasm for the practice of urban design. It also encouraged a broader passion for reflecting on patterns of urban development from a historical perspective. Working and living in Malaysia at this time, I grappled with many questions concerned with cultural identity and the extent to which entrenched sociocultural factors or practices affect the physical resolution of modern cities. Perhaps the vibrancy of life in my adopted city of Kuala Lumpur, the smells of sumptuous street food, the activity at all hours, the cultural diversity, the tropical environment ensuring many everyday practices occur in the public realm, and the practicalities and romance of the urban architecture led me to question motivations behind the choices people make in realizing lived space? I was intrigued by the shifting from what appeared to be a distinctly Southeast Asian style of modernization to prioritizing a Western style of urbanization. Southeast Asian cities, particularly in the 1990s, experienced wholesale expansion of their urban centers. In most instances, the fast-track urbanization of this region involved modern topographical analyses and two-dimensional master planning solutions, typically undertaken by imported design consultants. Professional practice made me aware of the common myriad factors contemporary cities are required to address. From the need for high-speed transport systems, to global aspirations for connectedness with communities beyond their immediate borders, to fostering local resources, identity, and talents, the complex nature of urban life seemed to be shaped by many factors. Yet why did rapid economic expansion of the late 1990s appear to translate the mechanical requirements of Southeast Asian cities into forms that seemingly reflected Western-inspired tower blocks and homogeneous urban environments? What are the most influential tools in the practice of urban design? How have humans imagined, designed, and realized their settlement throughout history? Have people always and everywhere thought about their cities in the same way? These and other related questions sparked my interest in the "map" and the notion of "mapping."

Why specifically look at the map? What role does the modern map play in the realization of an urban environment? The map is a cultural construct and seems to be an essential tool in the process of creating human environments. Although the complexities entrenched in mappings have been widely acknowledged by scholars of geography, cartography, chorography, anthropology, philosophy, and, to a degree, sociology, few studies have examined the evolution of the map and the city as synchronized events. To what extent has the map's changing nature affected Western and non-Western built contexts? And can the complexities of mapping reflect the meanings involved in the creation of human settlements? Are questions that have all received limited attention in current literature.

The primary aim of this study is to examine the fundamental connection between a visual representation of a form of human settlement, broadly referred to as a "map," and the reality of settlement itself. This examination is conducted in Western and non-Western contexts in order to understand, on the one hand, the agency of the map in spatial conception of settlement, and on the other, the meanings and narratives, direct or symbolic, that the map conveys. The study focuses on only one aspect amid many that influence the form of past and present cities: the historical evolution of the map and its effect on, and association with, the city and urban space from premodernity to postmodernity.

In this study, a "map" is understood primarily as a graphic interpretation of space, geography, or objects, using signs and symbols to give meaning to its chosen form of representation, which may even transcend the physical.[1] "Mapping" concerns the process of producing a map, in which data may consist of, but are not limited to, anything of a spiritual, imaginal, cosmological,

[1] Paul Laxton and J. B. Harley, *The New Nature of Maps: Essays in History of Cartography* (Baltimore: John Hopkins University Press, 2001), 16.

geographical, ideological, physical, or topographical nature, which is then, in a highly selective way, collected, collated, and interpreted into a type of representation of space. In modern scholarly studies, "mapping" has become a term widely adopted to describe a process of inquiry and, subsequently, the results of that process. The distinguishing factors of the mode of inquiry adopted here are, first, a focus on the sociocultural meanings of the map, and second, the possibilities of the agency of mapping for urban design and what maps reveal in representing different spatialities.[2] By following the historical role of the map and understanding its conceptual framework, we might be able to understand the ways in which maps and mapping affect design outcomes.

Human settlement, its patterns, influences, methods, and motivations, have been hypothesized for many a century, from both singular and multifaceted perspectives. Historically, social processes of settlement formation included place and society evolving together without any comprehensive preplanning undertaken, perhaps adapting to geographic difficulties or particular rules of land ownership.[3] Alternatives for urban formations included settlement patterns being controlled by certain rules dictated by social, religious, or legal understandings and requirements, as was the case with geomancy, the Laws of the Indies, the Islamic religious laws, or interpretations of literary works.[4] A major aspect of settlement formation can be traced to our cultural as well as our emotional needs

---

[2] James Corner, "The Agency of Mapping: Speculation, Critique and Invention," in *Mappings*, ed. Denis Cosgrove, 213–252 (London: Reaktion Books, 1999), 213. Corner differentiates between planning and mapping because the latter enables possibilities whereas planning limits potential in its tracing of space rather than creating it. See also, for the abstract and theoretical nature of maps, Denis Wood, *The Power of Maps* (New York: Guilford Press, 1992); Mark Monmonier, *How to Lie with Maps* (Chicago: University of Chicago Press, 1991); John Pickles, "Texts, Hermeneutics and Propaganda Maps," in *Writing Worlds*, eds. Trevor J. Barnes and James Duncan (London: Routledge, 1992).

[3] Tony Atkin and Joseph Rykwert, *Structure and Meaning in Human Settlements* (Philadelphia, PA: University of Pennsylvania Museum of Archaeology and Anthropology, 2005), 12.

[4] See Besim Selim Hakim, 1986. *Arabic-Islamic Cities: Building and Planning Principles*, 1st ed. (London: Kegan Paul International, original ed., 1979; repr., 1988); Besam Hakim, *Sidi Bou Said, Tunisia* (Halifax, Nova Scotia, Canada: Nova Scotia Technical College, 1978), 20–23, Call to prayer for this city meant residents of the old town located their houses within a 250-meter zone from the minaret to hear the voice of the muezzin. In addition, see Seyyed Hosein Nasr, *Ideals and Realities of Islam* (London: George Allen and Unwin Ltd, 1966). On Southeast Asia, where religion and religious traditions were seen as reasons behind certain indigenous built forms, see David Marr and Anthony Milner, *Southeast Asia in the 9th to 14th Centuries* (Singapore: Institute of Southeast Asian Studies, 1986): Elizabeth Bacus, Ian Glover, and Peter Sharrock, *Interpreting Southeast Asia's Past: Monuments, Image and Text* (Singapore: NUS Press, 2008).

and desires.[5] Some of these processes include consultation with the fields of cartography and geography, for collated data that explicate the origins of cities and settlement patterns.

Modernity has prompted a range of investigations into mapping processes in city creation, for example, works by 1960s art movements, such as the Dadaists, Surrealists, and later, architectural offices (e.g., Archigram and *Utopie*). These works, which were associated with the rise of the utopian movement, were included in new editions of Bannister Fletcher's book, recognized for its contribution as a source on the history of architecture and landscape.[6] These movements of modern inquiry interpreted different modes of mapping for creating built form, yet their practices were largely considered utopian and experimental and, therefore, on the whole, not credible options for realizing the city.[7] A contemporary resurgence in interest into investigations of an overlap between cartography and the creation of urban landscaped environments has occurred since protagonists of the landscape urbanism movement began adopting mapping processes for research and design in the late 1990s.[8] A flow-on effect resulted in the surface architecture genre of buildings, yet, like many earlier attempts, did not necessarily translate into applying these processes to analysis of the broader urban environment. Thus, in rethinking the agency of maps and the process of mapping, this study traces the sociocultural and intellectual contents of early maps and their influence historically, if any, on urban form and settlement patterns. Does settlement precede the map? On the other hand, does the map precede settlement?

---

[5] Atkin and Rykwert, *Structure and Meaning*, 9. The traditions and variety of Southeast Asian urban forms may be further understood through consulting texts such as Johannes Widodo, *The Boat in the City: Chinese Diaspora and the Architecture of Southeast Asian Cities* (Singapore: Marshall Cavendish Academic, 2004); also Daigoro Chihara, *Hindu-Buddhist Architecture in Southeast Asia* (Leiden: E. J. Brill, 1996).

[6] Reyner Banham, *Megastructure: Urban Futures of the Recent Past* (London: Thames and Hudson, 1976), 92. Refer to Chapter 5, "Fun and Flexibility," for a discussion of a myriad of projects in the mid-1960s that were characterized by their lack of "nut-and-bolt" proposals, yet elaborate drawings and models espoused their promotion of fantasy for new living environments, 84.

[7] Ibid., 101. On the use of Barbarella type images to populate these representations, Banham suggests, "nothing could more aptly illustrate the proposition that megastructures were ideal cities containing other people's Utopias."

[8] Mohsen Mostafavi, Ciro Najile, and Architectural Association, *Landscape Urbanism: A Manual for the Machinic Landscape* (London: Architectural Association, 2003), 7.

Moreover, to what extent does the map motivate and capture cultural influences of settlement? These are some of the questions this study attempts to address.

The study approaches the concepts of maps and mapping within the intellectual sphere developed in the writings of Denis Cosgrove, Thongchai Winichakul, Matthew Edney, and James Corner. Cosgrove argues, "mapping is a process which involves both a 'complex architecture of signs' … and a 'visual architecture' through which the worlds they construct are selected, translated, organized and shaped."[9] He highlights contemporary scholarly considerations of premodern maps, ranging from material objects to metaphors.[10] From a different standpoint, Winichakul examines the non-Western context of Siam (Thailand) in its resistance to colonization and the shaping of its identity and appreciation of nationhood, and the role of mapping in this process.[11] He acknowledges that people of different cultures mapped their worlds differently. This shows that there is something imaginational about mapping, and that "imagined geographies" are in essence the object of all mapping practices. Being highly imaginative, the premodern map, unlike its modern counterpart, does not reveal the same meanings universally to its viewers, but allows for interpretation of multiple meanings.[12] Winichakul states "to understand the space and the map, one has to understand its concept (its grammar) and its symbolism (its morphemes)."[13]

In Matthew Edney's study of the mapping and creation of colonial India, he asserts that modern understandings of geography cannot reconcile mapped space, which does not connect to the greater context of the earth, using latitude or longitude. By contrast, premodern mappings engaged and represented all aspects of life on earth, as well as life after death, in their conceptions of space. Colonials,

---

[9] Denis Cosgrove, "Introduction: Mapping Meaning," in Cosgrove, *Mappings*, 3.
[10] Ibid., 4.
[11] Thongchai Winichakul, *Siam Mapped: A History of the Geo-Body of a Nation* (Honolulu: University of Hawai'i Press, 1994), 13. See also Sumet Jumsai, *Naga: Cultural Origins of Siam and the West Pacific* (Singapore: Oxford University Press, 1988); Victor Lieberman, *Integration on the Mainland*, vol. 1 of *Strange Parallels: Southeast Asia in a Global Context, 800–1830* (Cambridge: Cambridge University Press, 2003). For an alternative non-Western case and point of view of the role of mapping in the fifteenth century in China, see Gavin Menzies, *1421: The Year China Discovered the World* (London: Bantam Books, 2002).
[12] Winichakul, Siam Mapped, 21. See also Paul Wheatley. 1961. The Golden Khersonese: Studies on Historical Geography of the Malay Peninsula before AD1500 (Kuala Lumpur: University of Malaya Press, 1961).
[13] Winichakul, *Siam Mapped*, 28.

in their cartographical enterprise in India, created a consolidated India in the eyes of the Indians, or posed as the "intellectual masters of the Indian landscape."[14] The trigonometrical survey of India conducted and drawn by the British, was the first graphic interpretation made available for viewing by Indian and British citizens to demonstrate the extents of the lands that signified the continent of India in its entirety. According to Edney, this cartographical process also occurred in other colonized territories, such as parts of British and Dutch Southeast Asia, whereby territory indicated on a map enabled the conception of a region.[15]

James Corner's work, with those of Charles Waldheim and Mohsen Mostafavi, in its review of the map in the contemporary context, has led to new disciplines of study and a new form of practice known as landscape urbanism.[16] Critically assessing notions of maps and mappings has revealed new ways of designing, in which mappings of data actively contribute to and sometimes participate in design outcomes in both landscape and urban designs, rather than being solely design tools. For Cosgrove and Corner, the process of mapping can act as a prism through which to view the changing nature of urban environments. Contemporary mappings are projected interactively into the design method. New types of spatialities are motivated around greater connectedness in urban environments physically and metaphorically, and then implemented through sociocultural practices.[17]

Unlike in premodern times, the map today is utilized predominantly as a documentary and design tool. It is used more as an "instrument" than as a "story," revealing the sociocultural and religious dispositions of the peoples that inhabit

---

[14] Matthew Edney, *Mapping an Empire: The Geographical Construction of British India 1765–1843* (Chicago: University of Chicago Press, 1997), 16. As a product of Enlightenment thinking with regard to correctness and accuracy, this spatiality of India was a creation of history and events. See also Paul Carter, *The Road to Botany Bay: A Spatial History* (London: Faber and Faber, 1987), xv.

[15] Edney, *Mapping an Empire*, 3. See also Donald K. Emmerson, "Southeast Asia: What's in a Name?" *Journal of Southeast Asian Studies* 15 (1984): 1–21. Until post-World War II, an understanding of the Southeast Asian region did not exist until it was able to be recognized in colonial maps. Elizabeth Bacus, Vincent Piggot, and Ian Glover, *Uncovering Southeast Asia's Past: Selected Papers from the 10th International Conference of the European Association of Southeast Asian Archaeologists* (Singapore: NUS Press, 2006).

[16] Mostafavi and Najle, *Landscape Urbanism*, 5.

[17] Cosgrove, "Introduction: Mapping Meaning," 5.

mapped space.[18] To what extent has this shift in the role and understanding of the map reflected similar changes in the evolution of our built environments? Historically, the creation of cities has occurred over vast tracts of time, responding to different technologies as they become available, and to changes in thinking about the way peoples should dwell and inhabit space. Typically, premodern cities grew organically, and mappings of their physical spaces and buildings occurred retrospectively. This chronology of events occurred for a variety of reasons, from invasion to colonization to creating registers of land holdings and entitlements. It is only in modern times that cities and their evolution have been completely dependent upon maps for the layout of key buildings, roads and other infrastructure, dwelling sectors and public spaces, which all occur prior to the creation of the city, and entirely in a premeditated and highly articulated fashion. Thus, gaining insights into the historical production of the map will provide us with clues to understanding the changing nature of the city's urban qualities.

Research Parameters: Space and Time

Space

In premodern times, maps were culturally specific. In an attempt to illustrate the cultural specificity of the premodern map and mapping processes, the study focuses on the example of Southeast Asia as a non-Western cultural context. For the purpose of this study, the region of Southeast Asia includes the following countries: Burma (Myanmar), Thailand, Laos, Cambodia, Malaysia, Philippines, Vietnam, Indonesia, Singapore, Brunei Darussalam, Timor Leste, and Malaysian Borneo. *The History of Cartography* volume on Southeast Asia excludes some of these countries in its definition of the region because of their cartographical influences being more closely aligned with countries, such as

---

[18] Ibid., 2. For the reductive nature of the modern map from the sociopolitical perspective, see James C. Scott, *Seeing Like a State: Why Certain Schemes to Improve the Human Condition Have Failed* (New Haven: Yale University Press, 1998). For mapping insight and the revealing potentials of the map, see Stephen Hall, *Mapping the Next Millenium: The Discovery of New Geographies* (New York: Random House, 1992).

China in the case of Vietnam.[19] However, in the context of this study, I do not support such exclusions because the countries within the region enjoyed many influences typically based upon religious belief systems, such as Theravada Buddhism, Islam, Hinduism, and Christianity. These influences most commonly came because of early trade and travelers from China, India, and the Middle East.[20] The region of Southeast Asia is a modern phenomenon and bears little significance as a concept or a conglomeration of territories for premodern Southeast Asian peoples, but rather served as a convenient way to identify the group of islands and different countries in the region in postcolonial times for the ruling hegemonies.[21]

In premodernity, Southeast Asia was not commonly understood as a region and group of countries, whilst concepts such as boundary and territory, also had quite different meanings between regional cultural groups.[22] Local peoples challenged western ideas about maps, such as indicating boundaries and territory, as these were foreign concepts to Southeast Asians, yet to Westerners they were the very basic knowledge and signs that constituted a map.[23] The notion of boundary in contemporary thinking maybe aligned with a permeable concept of premodernity, unlike early colonial conceptions.[24] Western assumptions about the form of the map were also undermined as archaeologists revealed temple architecture and other three-dimensional structures displaying cartographic signs to explain the passage of life from birth to death. Some examples included Hindu temples, which represented the first visual display of cosmographic ideas dating from the seventh century in Java. Overall, though, map-like images were still not

---

[19] David Woodward, "Preface," in *Cartography in the Traditional East and Southeast Asian Societies*, vol. 2, bk. 2, *The History of Cartography*, eds. J. B. Harley and David Woodward, xxiii-xxvii (Chicago: The University of Chicago Press, 1994), xxiv.

[20] See Anthony Reid, *Southeast Asia in the Age of Commerce* (USA: Yale University Press,Vol. 1Ithaca, 1988) for an exhaustive analysis of trade, economic activities which prompted increased urbanization in the region. Also for accounts of the role of the Chinese in European views of the world through maps, see Gavin Menzies, *1421: The Year a Magnificent Chinese Fleet Sailed to Italy and Ignited the Renaissance* (London: Harper Collins, 2008).

[21] Winichakul, *Siam Mapped*, 14.

[22] See Deborah Pellow, ed *Setting Boundaries: The Anthropology of Spatial and Social Organisation*. Conneticut, USA: Praeger, 1996) for an analysis of different understandings of the concept of boundary between indigenous peoples and Western interpretations.

[23] Winichakul, *Siam Mapped*, 68. Winichakul highlights many misunderstood examples of boundaries in his "clashes of conceptions of boundaries."

[24] Cosgrove, "Introduction: Mapping Meaning," 4.

common, only appearing in temples from the twelfth and thirteenth centuries. The frequency of appearances of temple mappings is not known.[25] Buddhist temples portraying similar information go back to the eighth century.[26] Although the medieval Christian world was highly cosmological, for Southeast Asian peoples the significance of space on earth was intrinsically linked to the afterworld. This meant that their ideological positioning motivated more musings about a connectedness to whatever lay beyond the earthly world than a purposeful investigation of further earthly territories. Mappings were in life and death sociocultural and religious representations of premodern Southeast Asian ideas. They explored ideas in cartographic media as well as other constructions, such as rock maps, verbal itineraries, stick charts, sand mandalas, and mythical stories. In many ways, premodern Southeast Asian mappings grappled with complex and timely transitioning between worlds of life and death, metaphorically and physically. Therefore, the breadth of territory to map may be considered more highly conjectural or ambitious than any European equivalent. Among the variety of these early attempts at understanding the world that Southeast Asians inhabited as well as the realm they would pass to after death, their immediate physical neighbors and associated territories were not viewed with the same urgency for attaining knowledge.

Time

Historically, the period of the study extends from premodernity to postmodernity. A few prehistoric examples and ideas are discussed because of their significance and contribution to the main aims of the study. Although premodernity in Southeast Asia may be said to have extended until colonization, colonial practices did not dominate local cartography until around the nineteenth century. Thus, the temporal borderline that separates pre- from early modernity is somewhat blurred. The sixteenth century corresponded with the great age of exploration for Western societies, and thus, European voyages prompted many changes in cartographical representations for indigenous and non-indigenous

---

[25] Joseph E. Schwartzberg, "Introduction to Southeast Asian Cartography," in Harley and Woodward, *Cartography in the Traditional East and Southeast Asian Societies*, 695.
[26] Ibid., 693, 695.

peoples.[27] Printing and copying processes reflected the progress of earlier exploration in interpretations of geographical and sociocultural qualities of non-European societies, discovered by sea in early modernity or late in the sixteenth and early seventeenth centuries. These altered ideas of foreign places filtered into the mapping genre, which encouraged mapmakers in their refining and sharing of information.[28] Although many Southeast Asian countries were colonized at different times, some, such as Siam, retained their sovereignty. The influence of Western practices in cartography began to filter throughout the region because of their perceived accuracy and sophistication of method.

Premodern Southeast Asians utilized different modes and approaches to convey ideas about space. Some examples were visual, such as cosmologies and sand *mandalas*. Other conceptualizations were verbal and in literary forms, and some were sculptural and three-dimensional in the form of sacred architecture. These methods remained largely unchanged until they gradually shifted and were reflected in cartographic works from post-nineteenth-century perspectives. Southeast Asia offers a geographical and temporal frame within which to differentiate between Western and non-Western mappings prior to the nineteenth century.[29]

The study also investigates significant changes that took place in early modernity in Europe because of modern science developments, which affected mapping and cartographic practices. In one seventeenth-century example, maps began to play a specific role in European societies, whose judiciary relied on them in land entitlement disputes.[30] In addition, as a graphic reference, they became significant in historical and secular knowledge, relating concepts of space to a

---

[27] Jeremy Black, *Maps and History: Constructing Images of the Past* (New Haven, London: Yale University Press, 2000), 6. Maps could be more easily reproduced and there were also more maps to refer to in creating revisions. See also Menzies, *1421*, 11, for an account on how China had made significant and accurate maps of new worlds prior to the Europeans in the mid-fifteenth century.

[28] Black, *Maps and History*, 6.

[29] Woodward, "Preface," xxiv.

[30] Black, *Maps and History*, 6. See also D. H. Fletcher, *The Emergence of Estate Maps. Christ Church, Oxford c. 1600–1840* (Oxford: Oxford University Press, 1995).

place, while also offering the possibility of demonstrating space, historically, in an objective manner.[31]

The seventeenth century also signified changes in the relations of time and space that had implications for the reading and understanding of both geography and history. As conceptualizations of space changed in Europe because of scientific innovations in optics, measurement of longitude and latitude, triangulation techniques, and the measurement of time, techniques of modern mapping also evolved. These developments were not unilaterally transferred to non-European nations. However, seventeenth century Dutch cartography was clearly affected, for example, by requiring a greater degree of realism or topographical accuracy to be achieved by the mapmaker.[32]

Although the focus of the study is on premodern and early modern developments in mapping, it also investigates how these changes and innovations in mapping affected conceptualizations of urban environments that resonated into the twentieth century and to current practices. For example, the modern movement adopted the map developed from Enlightenment technologies as a source or a tool with which to rationalize space. The mantra of "form follows function," synonymous with modernism, supported these principles into the design process, and ultimately into the created environment.[33] By contrast, late-twentieth-century practices, such as landscape urbanism, shifted to prioritize mapping as a process, denying the map its previous stasis and rationalism, focusing on flows of information and the dynamic nature of space, and

[31] Black, *Maps and History*, 7. This was also due to the influence of humanism and literal interpretations of texts over the allegorical aspects of religious writings. See also P. Burke, *The Renaissance Sense of the Past* (London: Edward Arnold, 1969).

[32] Black, *Maps and History*, 8. See also S. Alpers, *The Art of Describing: Dutch Art in the Seventeenth Century* (Chicago: The University of Chicago Press, 1983), chapter "The Mapping Impulse in Dutch Art."

[33] "Form follows function," although adopted as a phrase by the modernists was first mentioned by American sculptor Horatio Greenough, who in 1852 was relating it to the organic principles of architecture. The American architect Louis Sullivan, who admired rationalist thinkers such as Greenough, Thoreau, Emerson, Whitman, and Melville, used it, in 1896, in his article "The Tall Office Building Artistically Considered." Here Sullivan actually said "form ever follows function," but the simpler (and less emphatic) phrase is the one usually remembered. Louis Sullivan's phrase "form (ever) follows function" became a battle cry of modernist architects after the 1930s. Form follows function implied decorative elements, or any type of external feature or "ornament," were superfluous in modern buildings. "Form follows function" accessed March 20, 2012, http://en.wikipedia.org/wiki/Form_follows_function.

questioning how that might be reinterpreted in the creation of contemporary urbanities.

Method

Sources

The methodology of the study is based on archival research, examining textual and visual materials. The topic requires selection, collating, and examination of sources from interdisciplinary fields, including geography, cartography, social science, anthropology, urban studies, urban design, urban history and theories, architecture, and landscape architecture, all of which influence mapping practices and processes. This approach highlights the intrinsically complex characteristics of the map, and its interdisciplinary nature, while shifting the conventional focus from its technical properties to its sociocultural, religious, and intellectual qualities.

The primary sources used in this study are both historical and theoretical. They include the *History of Cartography*, of which the Southeast Asian and East Asian volumes are consulted for the most part.[34] *Early Modern Mapping in Southeast Asia* by Thomas Suárez is included because it features some indigenous maps that have not been previously published and were sourced from private collections.[35] In addition, Ralph Ehrenberg's pictorial *Mapping the World* presents an insightful graphic study of some key European maps.[36] All sources are in English and by predominantly Western-educated authors. Although some indigenous materials are included, this study is limited to works that have been published in the English language. Each of the sources has been selected for its insight and influence within its given field, and its use of original archival material. Where appropriate, I have attempted to identify and acknowledge biases and weaknesses of each source, especially with regard to dealing with original indigenous sources.

The selected studies on Southeast Asian urban history and mapping, such as *The History of Cartography*, as well as *Early Mapping of Southeast Asia*, are

---

[34] Harley and Woodward, *Cartography in the Traditional East and Southeast Asian Societies.*
[35] Thomas Suárez, *Early Mapping of Southeast Asia* (Singapore: Periplus Editions (HK), 1999).
[36] Ralph E. Ehrenberg, *Mapping the World: An Illustrated History of Cartography* (Washington, DC: National Geographic Society, 2006).

considered exemplary in their collation of descriptions and interpretations of indigenous and non-indigenous mappings, thereby enabling access to premodern and early modern maps. They identified the extent of the surviving corpus of published Southeast Asian maps, and outlined the difficulties of studying historical material of another culture and time, as well as problems with interpreting it within the realm of our modern conceptions of space and maps. Mapping material was sometimes prepared by indigenous peoples at the request of Europeans, as in the eighteenth-century case of the Burmese assisting an Englishman named Hamilton.[37] There were also other problems of storage and preservation of archival information, for example, in Thailand, where the concept of cleansing and purification (*Chamra*) governed archival management and has affected access to original copies. *Chamra* has a long history and is considered an important practice in Thai culture.[38] It may refer to a mild alteration of the document, such as correcting grammatical errors, modernizing the spelling, or updating titles of individuals.[39] More radical alterations include filling in blank areas of manuscripts, rewriting of materials, and removal of some documents from circulation. *Chamra* is a continuous process and as ideas change over time so does *Chamra*. Therefore, Thai manuscripts cannot be viewed as stable entities.[40] The difficulties involved in this mapping practice range from determining authorship and authenticity of original material to interpretation of premodern cartographical signs.

Secondary sources for this study include texts that offer insights on critical interpretations of mapping, maps and cartographic, as well as urban history. These groups of texts have provided a theoretical and analytical frame through which to read the primary sources. They have also enabled an informed understanding of how the changing role of the map through early modern history has critically affected the creation of human urban habitats and ideas about settlement. To this

---

[37] Schwartzberg, "Introduction to Southeast Asian Cartography," 699. See also R. T. Fell, *Early Maps of South-East Asia* (Singapore: Oxford University Press, 1988) for the European perspective on Southeast Asian mappings. In addition, a study by Reginald Henry Phillimore, "An Early Map of the Malay Peninsula," *Imago Mundi* 13 (1956): 174–79.

[38] Schwartzberg, "Introduction to Southeast Asian Cartography," 698. See also Constance Wilson, "Cultural Values and Record Keeping in Thailand," *CORMOSEA [Committee on Research Materials on Southeast Asia] Bulletin* 10 no. 2 (1982): 4–6.

[39] Schwartzberg, "Introduction to Southeast Asian Cartography," 698.

[40] Ibid.

end, authors such as Denis Cosgrove, James Corner, and Françoise Choay have enabled the positioning of this study to analyze the data and material provided in its historical body.

Denis Cosgrove's edited book *Mappings* entails a series of essays, highlighting contemporary thinking and critical analyses around the changing nature of the map.[41] Cosgrove expands on the initial definitive work of Brian Harley, who challenged the triumphalist nature as well as assumptions about the accuracy of information the map might portray.[42] Cosgrove accords contemporary interest in mapping processes and the map to the rapidly changing nature of postmodern society compared with that of modernity. Twenty-first-century culture connects closely and often simultaneously via data networks with the global sphere, and thus the fixing of space to certain boundaries has taken on different meanings in today's society. There is now a greater focus on theorizing space from dynamic and interconnected perspectives. Deleuze and Guattari view space from a sociopolitical perspective associated as highly structured and "arborescent," or of the "rhizome type" that is neither structured nor hierarchical but more sporadic and unplanned in reference to the nature of rhyzomic connections and movements of contemporary society.[43] In summary, Cosgrove and others question the very assumptions upon which mappings have typically been based and any implied stasis associated with space, which may be mapped, conceived, and therefore understood.

James Corner speaks of the "agency of mapping" and acknowledges the benefits of such a process in reshaping ideas about mapping as well as the map itself.[44] In this case, Corner does not dwell on the power relations of the map but

---

[41] Cosgrove, *Mappings*. See also Roger Starling, "Rethinking the Power of Maps: Some Reflections on Paper Landscapes," *Ecumene* 5 (1998): 105–8; Ola Söderstrom, "Paper Cities: Visual Thinking in Urban Planning," *Ecumene* 3 (1996): 249–81.

[42] J. B. Harley, "Maps, Knowledge and Power," in *The Iconography of Landscape: Essays on the Symbolic Representation, Design and Use of Past Environments*, eds. Denis Cosgrove and Stephen J. Daniels, 277–312 (Cambridge: Cambridge University Press, 1987), 278. Opposing the binary oppositions set up in mapping, i.e. accurate and inaccurate, objective and subjective, Harley highlights that maps always have sets of values. See also to compare Janet Wolff, *The Social Production of Art* (New York: New York University Press, 1981), chapter "Art As Ideology."

[43] Gilles Deleuze and Felix Guattari, *Nomadology: The War Machine*, trans. Brian Massumi (New York: University of Minnesota Press, 1986), 13.

[44] Corner, "Agency of Mapping," 213. See also James Corner, *Recovering Landscape: Essays in Contemporary Landscape Theory* (New York: Princeton Architectural Press, 1999), chapter "Operational Eidetics in Forging New Landscapes."

rather adopts its liberating capacity for designers of urban space.[45] The practice of mapping enables the remaking of space and territories, each time the space is recreated through a mapping. A copied map or what Corner calls "tracing" does not permit such possibilities.[46] Corner suggests that maps are "analogous-abstract" in nature and offer up both the calculated surface of the geographic context in question, which is subsequently projected onto the graphic representation, and an abstract quality of context—that is the information that is not relayed via the mapping.[47] It is inherently a selection process; therefore, a degree of abstractedness results. Therefore, Corner states his focus is not on the finished, complete product of the map, or its meaning, but rather on what the product does to or for the design creative process.[48] I explore this perspective in my study, and the final chapter explores some contemporary practitioners in this area.

Françoise Choay's work *The Rule and the Model* focuses upon written theoretical foundations of concepts of the urban environment.[49] Although written accounts of human environments and ideas of space date back most notably to ancient Chinese texts, Choay's focusing on the theoretical argues that the "creation of an autonomous discourse" on theories of spatial configurations is a Western preoccupation.[50] Choay highlights that, in non-Western contexts and in societies without writing, "the organization of built space is interconnected with the whole of social practices and representations, without there having to be a word in the lexicon dedicated to reflection on the idea of spatial arrangement."[51] Choay attributes the origins of urban theory in Western discourse to the first architectural treatise of the fifteenth century, namely, Leon Battista Alberti's *De re aedificatoria*.[52] The treatise identifies a set of rational rules, with its basis in

[45] See James Corner, "Representation and Landscape," *Word & Image* VIII, no. 3 (1992): 243–75.
[46] Corner, "Agency of Mapping," 214. See also Gilles Deleuze and Felix Guattari, *A Thousand Plateaus: Capitalism and Schizophrenia*, trans. Brian Massumi (Minneapolis: University of Minnesota Press, 1987), 12.
[47] Corner, "Agency of Mapping," 215. See also J. B. Harley, "Deconstructing the Map," in Barnes and Duncan, *Writing Worlds*; Harley, "Maps, Knowledge and Power."
[48] Corner, "Agency of Mapping," 217.
[49] Françoise Choay, *The Rule and the Model*, ed. Denise Bratton (1980; repr., Cambridge, MA: Massachusetts Institute of Technology, 1997).
[50] Ibid., 3. Whereby the term "urbanism" was coined in 1867 by Ildefonso Cerda.
[51] Ibid.
[52] Ibid.

developments of mathematics and physics of the period, with the "explicit aim of developing an autonomous conceptual apparatus in order to conceive and build new and unknown forms of space."[53] It is in the authorship of this text that Choay identifies, the period in the Western history of urbanism, where a shift from non-theorized contexts to theorizing text in relation to the built environment occurs. In effect, *De re aedificatoria* established another means by which to conceptualize, draw, or map one's world.

The third groups of sources are studies that are close to aspects of my work, yet also different, include, Matthew Edney, Thongchai Winichakul, and again, James Corner. The latter author is mentioned twice because this study adopts his critical perspective of mapping in urban history and urban design. Matthew Edney's major work *Mapping the Empire* investigates the Great Trigonometrical Survey of India.[54] Although the work is about the British colonization of India, the argument presented regularly refers to the Southeast Asian context. This is for comparison and to demonstrate what the British were doing with their evaluation of geographical data in their other colonies, such as Malaysia and Singapore.[55] Edney and it offers insights into the technical strategies that were available to colonizers for achieving their geographical visions.[56] The power of the map is asserted, because of these studies, as a great tool of psychology. For the Southeast Asian condition, although earthly geography was not extensive and connected in mappings, indigenous thoughts beyond the immediate realm of the village did most convincingly extend to greater realms of spatial considerations. Cosmological considerations extended spatial dimensions and understandings of the world beyond. Southeast Asian concepts of physical space may not have aligned with Western premodern understandings, yet cosmological and temple mappings seem to suggest that indigenous spatialities

---

[53] Ibid., 6.

[54] Edney, *Mapping an Empire*. See also Gwendolyn Wright, *The Politics of Design in French Colonial Urbanism* (Chicago: The University of Chicago Press, 1991).

[55] Note: British Malaya (18th century–1946), a loose collection of British-controlled states. Modern term, post-independence Malaysia.

[56] Benedict Anderson, *Imagined Communities: Reflections on the Origin and Spread of Nationalism*, revised 2nd ed. (London, New York: Verso, 1991). Anderson suggests that both Thailand and Indonesia seemed to adopt imaginings with regard to their countries prior to the colonial era, 163, 174–78.

were typically both projective and retrospective spaces, rather than just a translation of a physical place.

Another source that is substantially referenced in this study is the work of the Western-educated Thai scholar, Thongchai Winichakul, *Siam Mapped: A History of the Geo-Body of a Nation.*[57] In this work, one-third of the references are translated from original Thai sources. Yet one of the most valuable contributions of the work is the idea of "geo-body" and the way it is constructed to permit a legitimate and viable retrospective and projective analysis of the subject matter. This proposition allows the study of the history of Thailand as a theoretical concept and as a given territorial entity, using the author's notion of geo-body. Winichakul does not present a linear history but prioritizes other understandings of space, proposing it as a guide to consider premodern Thai history. In effect, the geo-body of Siam reflects a geographical area on a map as well as the characteristics that make up the "life of a nation."[58] It is this very understanding of the complexity of premodern versus modern interpretations of space that make this text useful for my study. Winichakul highlights the dilemma of whether territory exists before or sometimes only on the map.[59] For this reason, it is important to open up our conceptions of mappings because they are not always focused on territory *per se.* Similarly, Winichakul champions Mongkut, who was part of the Siamese aristocracy (he later became King Rama IV, r. 1851–1868), for his accurate predictions of geography and topography based on indigenous astrology and modern astronomy.[60] These interpretations of geography apply to the specificity of the Thai case because it remained independent in the colonial era. Unlike those in colonized countries, Siamese mapmakers had freedom to choose to follow traditional or modern methods or combinations of the two in their spatial interpretations. Winichakul's text is highly persuasive in its delivery of the idea that, whatever the achievements of modernity, most studies on non-Western contexts appropriated historical cultures and misrepresented them according to their own subjective views and interests. The acceptance of a degree of bias in any composition is an important aspect in the exchange of human

---

[57] Winichakul, *Siam Mapped.*
[58] Ibid., 17.
[59] Ibid., 56. Edney concurs with this idea in the context of India.
[60] Ibid., 37.

knowledge, and this study does not claim to be free from such bias in the views of every reader.

Limitations and Contributions

This is an interdisciplinary study that bridges three major fields: urban history, urban design, and the history of cartography. The three fields are large and their literatures are extensive; consequently, many limitations confine this study. In addition, the study utilizes insights from other disciplines, most notably geography, anthropology, and architectural and landscape architectural history, to trace other facets of meanings in settlement formation. Thus, the study can only deal with a small sample of selected literature from each field to construct the thesis and sustain the core arguments. Yet, it is in this interdisciplinary research that the main contributions of the book lie. This study derives insights from many strands of literature, across a multifaceted field of research, bringing them together to bear on urban design issues. The cross-sectional depiction of human settlement, viewed through the lens of mapping over a long historical span and in three major fields, has enabled this study to present fresh insights into the field of urban design, which rarely engages in such protracted historical and theoretical explorations. By examining the role of the map at a conjunction of urban history, urban design, and history of cartography, the study is able to show how people in medieval Europe and Southeast Asia conceived of and mapped their spaces of habitation, how the European Enlightenment's rational mapping proliferated into the non-Western world, how the production of urban space shifted from a socioculturally motivated style to a highly theorized framework, how the concept of the modern city was born alongside the emergence of modern urban planning, how the emergence of modern thinking about the city corresponded with new ways of designing, and how theorists reacted to the modernist urban design rationalism, which was anchored in the authority of scientific mapping. Through this new path of inquiry, the study is also able to uncover some of the lost meanings and functions of the map, and to examine new approaches to dealing with the loss of quality and identity in today's urban environments. It is in the construction of this narrative from a multifaceted, historically based theoretical perspective that the study delivers its strands of contributions.

Such an interdisciplinary study involves some inevitable conceptual and methodological problems and challenges. Southeast Asia has only been recognized as a region since post-World War II; therefore, the grouping of the countries in the region is a modern phenomenon, which inherently creates a conflict between modern conceptions and premodern conditions.[61] Thus, the task of selecting sources on Southeast Asia as the most authentic poses some challenges. There are also difficulties associated with the research as far as making generalizations owing to the grouping of different sociocultural contexts into regions, for example, not only in the case of Southeast Asia but also for the European conditions. Sources utilized in this study are predominantly written in English by Western authors; thus, it is legitimate to question their reliability in representing indigenous views.[62] Accessing original sources proved difficult predominantly because of language limitations, for which a working knowledge of Thai, Bahasa, Mandarin, Tagalog, Mandarin, Vietnamese, Lao, and/or Hindi is required. The scope of the work has thus been restricted to the sources mentioned above. However, these limitations are only relevant to the first part of the book, which examines historical texts dealing with the Southeast Asian condition and its mapping history. As regards this aspect of the study, the main contributions this study makes lie not so much in the historical material itself, but in the ways in which these historical insights are made relevant to, and brought to bear on, contemporary practices. They lie in the ways in which urban history is read together with urban design, an interdisciplinary reading that allows the conventional understanding of maps and mapping to be seen in a new light. In tracing the role of the map through the Enlightenment and the significant changes

---

[61] See Andrew Russel Wallace, *The Malay Archipelago* (1869; New York: Dover Publications, 1962), for the term "Malesia" (and later Malaysia 1963), based on bio-geographical references for flora and fauna studies based in the Malay archipelago. See also Peter Bellwood, *Prehistory of the Indo-Malaysian Archipelago* (1985; repr., Honolulu: University of Hawai'i Press Original edition, 1997) for the term of prehistory "the Indo-Malaysian Archipelago."

[62] See Lai Chee Kien, "Southeast Asian Spatial Histories and Historiographies," *The Journal of the Society of Architectural Historians, Australia and New Zealand* 19, no. 2 (2010): 82–105, who presents an interesting overview of Western-based and English sources, from the point of view of spatial construction in Southeast Asia. See also Kuo-wei Chen, "Meaning in Architectural and Urban Space of the Penang Kongsi Enclave" (paper presented at The Penang Story—International Conference, Penang Malaysia, 18-21 April, 2002.

associated with it during this period, the study provides a new lens to re-view the role of the map as an instrument for design purposes.

Book Layout

The book begins with the historical collation of information followed by a critical analysis of the findings, summing up with some alternative and contemporary practices, which demonstrate how notions of maps and mapping are being revisited in an urban design capacity in early twenty-first-century practices. The book structure consists of three main parts. Part I: Maps and Settlements examines how and in what form premodern mapping existed and survived; Part II: Mapping and Modernity identifies the key influences upon mapping from the Enlightenment to modernity; Part III: Mapping and Modern Urbanism examines some examples of mapping in urban design from the rationalist approach utilized for the creation of twentieth-century modernist designs. It also examines some examples of contemporary mapping practices and movements that ultimately arose as reactions to the modern movement.

Part I utilizes each of its three chapters to investigate the breadth of maps of premodernity, which embraced both cosmological and cartographical concerns. Chapter One reflects upon various ways that premodern cultures conceived their world through drawing, and how these drawings differed from mappings in their embrace of certain cartographic signs. While revealing and discussing the idiosyncrasies of premodern conceptions of space linked to sacred and profane geographies, this chapter, in turn, demonstrates the interconnected nature of premodern understandings of cartography and cosmography. Extrapolating on these ideas of representation, Chapter Two identifies five different themes of mappings typical to premodern and non-Western worlds. Chapter Three explores mapping in its relationship to premodern ideas about settlement. Focusing on Southeast Asian urban conditions, tracing evolution of settlements also uncovers the complexities in reading archaeological findings retrospectively. This dilemma is further elaborated in contemplating how premodern geographies and settlements were represented, many of which were combinations of sacred and earthly spaces.

To illuminate the significance of the transition, from conceiving settlement for sacred and sociocultural purposes, to our modern scientific understanding, Part

II, and first, Chapter Four, focuses on Enlightenment developments. The European Enlightenment signified shifts in mapping practices due to developments in cartography, geography, and other sciences that made way for our modern way of seeing the world. This shift in conceptualization also paralleled the Second Great Age of Exploration whereby first hand encounters produced conflict with previously assumed religious and geographic knowledge. Religious and scientific thinking became incompatible because the geographical archive was prepared by European powers for colonial motivations. Colonial locations including surrounding areas in Southeast Asia provided opportunities to put Enlightenment thinking into practice. New ideas and developments in mapping and settlement design stemming from the rationalized and reasoned Enlightenment processes were implanted into new colonial territories. A context removed from entrenched European sociocultural and religious considerations perhaps permitted a less contentious transition to modern settlement formation. Southeast Asians' sacred and profane conceptions of space were misunderstood by Europeans and this lack of understanding facilitated a general disregard for indigenous conceptions of space. In addition, local peoples presented little effective physical and/or psychological opposition to the new colonial hierarchies. The emergence of modern mapping was more easily projected and enacted because of colonization around the eighteenth and nineteenth centuries. However, Chapter Five highlights how the disjunction between colonial ideas of mapping and the indigenous views of space were exacerbated in maps, while also noting the lack of immediate traction of Enlightenment technologies to be adopted in the colonies. In addition to the problems associated with mapping new lands, there were the cross-cultural misunderstandings of mapping practices and its purpose.

Finally, Part III of the book, consisting of Chapters Six and Seven, expands on the Enlightenment project and the results of ultimate rationalized spatial understandings that ignited utopian gestures of the modern movement. Again, technological developments, but this time due to the Industrial Revolution, prompted new ways of thinking about modes of production of human environments. Mapping was instrumental in this new mode of production because it relayed rationalized space and conceptual principles clearly and concisely in representations of the survey and master plan. Yet the modern movement was deemed by critics to have failed in its delivery of its utopian ideas around the overthrow of the woes of the industrial city. Thus, the shift from an intrinsic relationship between meaning instilled in sociocultural space to a projective, supplanted, highly planned and designed spatial realization proved problematic. Chapter Seven then evaluates some movements, ranging from the Situationists to contemporary practitioners of ecological urbanism, to show how they too have revisited maps and mapping processes to provide other methods for achieving projective design.[63] Many of these movements embrace the potentials of mapping processes with aims to use the agency of the map for connecting the fields of urban history and urban design. It is the intention of this interdisciplinary study that it act as a starting point, in its overview of mapping processes, and aid future theoretical investigations into these fields of inquiry.

---

[63] Abidin Kusno, "Imagining Regionalism, Re-Fashioning Orientalism: Some Current Architectural Discourses in Southeast Asia," *Journal of Southeast Asian Architecture* 4 (2000): 45–61. See Kusno for critical assessment of Southeast Asian understanding of region. He examines the theoretical positioning of Banister Fletcher, Spiro Kostof, Sumet Jumsai, Tay Kheng Soon and Ken Yeang in the debate on regionalism as a reaction against modernism.

PART I: MAPS AND SETTLEMENTS:

An Historical Overview

CHAPTER ONE
Premodern Mapping

From ancient civilizations to premodernity, graphically representing one's world has taken many forms. It has not always been clear what a map entails and what distinguishes a drawing. It is only in modern times that maps clarified such ambiguities with codes and techniques to attempt universal understandings. In premodernity, maps were highly integrated with earthly and heavenly narratives or depictions. Maps constructed geographies for purposes ranging from travel, pilgrimage, and military encounters to understanding the passage of life and death. Premodern geographies were preoccupied with geopolitical and georeligious concerns and practices, which sustained their imaginational nature of representation. In the Christian tradition, placing biblical stories in a geographic context permitted comprehension of an actual history—an event in a time and place. Mapping harnessed the imagination, graphically correlating cosmographical and religious ideas. Mapping presented a physical reality of the sacred and profane by visually translating it into a picture plane. Premodernity rendered earthly and heavenly space inseparable and often indistinguishable on maps.

1.1 Drawing One's World

A fascination for drawing the world, representing it graphically, or conjuring its visual or textual "image" has characterized civilizations for millennia. Yet "drawing" remains a complex concept with more than sixty definitions for the word "drawing" in the English language alone.[64] The same can be said about the terms "map" and "mapping," whose meanings are closely connected with those of drawing. Generally, drawing means marking or representing, using lines, figures, symbols, or other graphic elements. When culturally coded, these marks or drawings convey shared meanings visually through an understood correspondence with what is being drawn. The "map" emerges out of culturally coded visual correspondence that loads a particular drawing with information about one's world, be it natural or constructed. This process of cultural codification can be seen as the only defining difference between what may be considered a map and a drawing. Drawing also means delineating spaces through the construction of an image, depicting or representing, recording, and framing or composing (in text or pictures). "I thought it necessary to *drawe* a *treatyse* for myself," is a quotation from a sixteenth-century pilgrim that shows the close connection between drawing and writing in making sense of one's world.[65]

Our modern conception of a geographical map is one that is tied to the use of techniques such as scale, orientation, and a variety of forms of measurement and delineation of topographical features, such as ground level relative to sea level and distance between geographical locations. All of these techniques contribute to our modern understanding of space and are demonstrated in drawing rather than writing. They are also techniques intrinsically linked to our modern conceptions of space; it is difficult to understand or relate to spatial representations in their absence. Reading a map and drawing to scale are taken-for-granted activities by many today. Basic concepts of scale, such as proportional reduction of the size of an object to be portrayed on paper at, for example, one hundred or fifty times smaller (1:100, 1:50), are widely shared conventions across many fields. Also widely practiced is the use of the scale ruler, whereby dimensions are indicated as a translation between reality and its scaled equivalent. Using the 1:100 scale side of a ruler, for example, one can easily work out that twenty meters measure twenty millimeters on the ruler with no calculation required.

---

[64] J. A. Simpson and E. S. C. Weiner. *The Oxford English Dictionary*, in *The Compact Oxford English Dictionary*, second ed., s.v. "Drawing."
[65] Ibid.

-The act of mapping in our contemporary context has also taken on a very broad perspective. The term, "mapping" is one means of representing or charting a large collation of data for analysis. The mapping process is significant in that the resultant product, be it a chart, table, diagram, or map, acts as an *objective* tool in framing and guiding decision making. The data collated do not need to relate solely to elements of space, but can reference a vast array of subjects, from body parts to genealogy of a species. In addition, modern spatial maps have also diversified in their ability to convey spatial relations from fixed, two-dimensional representations to the multidimensional computer-animated models. However, the map remains a form of codified abstract representation that organizes, structures, legitimizes, simplifies, highlights, constructs, and reevaluates knowledge and assumptions about the information being mapped. In one form, it can act as an important tool for visualizing the degrees of complexity in the terrain of any given geographical location, yet, in another, it can offer insights into sociocultural practices of a region through the use of vignettes or statistical data and records.

Mapping to convey spatial relations—the intended product being a map—is to give an understanding of the target context's extents, highlight significant features, and propose limits, which may also reveal the emotive nature of a place as well as its physicality. The format of contemporary mapped information varies between that which is intended for mass production to independent commissions, or maps for a specific purpose, from thematic maps to three-dimensionally constructed space and computer-generated views. Often, data collection of physical characteristics of environments results in two-dimensional interpretations. Specifically designed for consumer markets, such mapped space is an understood and expected phenomenon within contemporary society. Increasingly, though, with the advent of three-dimensional technology for the mass market, such as Google Earth, it is now inexpensive and convenient to attain three-dimensional projections.

Maps can be used to construct routes through the complex amounts of information that are now available to us so that we can evaluate in a structured way, make sense of, read, and respond to our contemporary living environments. By contrast, drawing does not necessarily structure information using the same codification principles of mapping to convey a message. Drawings are permitted to be more abstract, offering multiple readings that are meaningful among cultural groups. Although maps can offer some of these attributes, not all maps do. The common topographical survey, for example, offers only a limited and select spectrum of technical information.

Unlike premodern maps, modern maps are consulted authoritatively, given their depiction of space to scale, to a compass orientation, as well as their inclusion of details and measurements of space. Codification systems of modern mapping act without the additional support of written treatises, but rely on a legend or key as well as common and assumed understandings of cartographic signs. The distinction between drawing and writing in the premodern world was not as precise. Historically, words often essentially accompanied drawings and maps, or in the case of ancient Greek architects, such as Ictinus and Kallicrates, buildings and urban spaces were constructed via "detailed descriptive specifications which included dimensions," independently of drawings per se.[66]

The following discussion traces the evolution of drawings in the forms of maps from the earliest known example, and examines the notion of "drawing" in the context of texts and records to address questions such as: How did different communities and groups view or imagine the world and record their observations visually? What did they see and map? Can one civilization be differentiated from another by its established drawing or mapping practices, and how?

Drawings as Maps

According to Ralph Ehrenberg in "Mapping the World," premodern mapping practices were evident in three main geographical world regions—Asia, the Middle East, and Europe—from the seventh century BC to the end of the fifteenth century AD. Broadly speaking, premodern mapping practices were specific to these regions until around the end of the fifteenth century, when there was a heightening of information exchanges among peoples of different civilizations. Aesthetically, maps of these regions may be differentiated by their styles and artistic qualities, such as those of Chinese landscape drawings,

---

[66] Mark Hewitt, "Representational Forms and Modes of Conception: An Approach to the History of Architectural Drawing," *Journal of Architectural Education* 39, no. 2 (1985): 2. See also J. Coulton, *Ancient Greek Architects at Work* (Ithaca, NY: Cornell University Press, 1977).

Islamic miniatures, or European illumination.[67] Orientation was a key difference between cultures and regions. Early Christians, and thus European maps, oriented their views east toward Jerusalem, which was typical of *mappaemundi*. The Islamic mapmakers opted instead for a focus on the south, which was the direction of Mecca and Medina. Portolan charts did not always indicate orientation, whereas Asian accounts typically indicated south at the top of the page.

One of the oldest known world maps dates back to a Babylonian clay tablet example from the sixth century BC, and writing, drawing, and surveying are associated with both Babylon and Egypt as early as the third millennium BC.[68] The Babylonians are renowned for developments and practices such as constructing irrigation canals, surveying property lines, and representing their urban communities. In addition, the Babylonians utilized key cartographic concepts—scale and orientation—to convey a sense of their physical surroundings, the parameters of their urban environments, and the cosmological beliefs that linked the physical with the metaphysical.[69] Egyptian drawings of a similar period were characterized by embellished portrayals of everyday life, together with a crossover into the imaginary.[70]

Asian mapping traditions embodied, in a manner similar to the Islamic tradition, functional and scientific ideas represented in the *Yu ji tu*, a remarkably accurate map of China from the twelfth century. In addition, there are spiritually influenced maps, such as a 1364 Buddhist pilgrimage map of the "Five Indies," that display religious influences in the region.[71]

Paper was invented in China in the second century AD but was not used in Europe as a mapmaking medium until the eleventh century. Representation practices typical to Asian mapmaking at this time, included painting, calligraphy, and rubbings. Rubbings were the earliest form of copying or mass-producing of images. Therefore, with the invention of paper, images could be transferred from a master document, typically stone etchings, by rubbing the image on the

---

[67] Ehrenberg, *Mapping the World*, 17.
[68] Ibid., 21.
[69] Ibid.
[70] Ibid., 9.
[71] Ibid., 17. See also Robert R. Temple, *The Genius of China: 3000 Years of Science, Discovery and Invention* (London: Prion, 1998).

transportable medium of paper. Other media utilized by Asian mapmakers to portray their representation include animal hides and silk. Chinese landscape painters undertook beautiful mapped renditions of space, and a separate specialty trade emerged for the creation of Chinese sea charts.[72]

Ancient Chinese representations of the world ranged from incorporating map elements, such as direction, distance, and terrain, but utilizing them through text in accompaniment of the drawing, to mural paintings, capturing a sense of space through the artistic conventions of landscape painters of the Qin and Han dynasties (221 BC–220 AD). The purpose of these Chinese works ranged from purely artistic appreciation and spiritual connection to a place, to military as well as strategic political policies.[73] Yet, such examples as that of the Han Dynasty demonstrate the close connection between maps and paintings of this period. The use of variable perspectives to relay detailed scenes is not uncommon in other known archival examples.[74] In the fourth century BC, Chinese documents recall the use of the "south point" as the preferred direction. The early Chinese compasses were constructed of lodestone, which had magnetic properties to attract iron, consistent with feng shui beliefs, and used as tools for orientating buildings and graves. Although Chinese sailors were orienting their sea voyages by the south point in the thirteenth century, it was not until the end of the thirteenth century that the compass was perfected by the Italians.[75]

Western cartographical constructions of space are largely based on ancient Greek concepts dated from the fifth century BC, such as Pythagorean notions of a spherical earth. By the second century AD, Greco-Egyptian geographer and astronomer Claudius Ptolemy had collated theories of earlier thinkers to produce works that influenced future representations of the world for ten or more centuries. However, it was the Romans who focused on methods of practically recording aspects of physical geography and topography, as well as the art of

---

[72] Ehrenberg, *Mapping the World*, 17. See also William Reynolds Beal Acker, trans., *Some T'ang and Pre T'ang Texts on Chinese Painting* (Leiden: E. J. Brill, 1954–1974).
[73] Ehrenberg, *Mapping the World*, 9.
[74] Harley and Woodward, *History of Cartography*, 147. For a detailed discussion on the connections between Chinese painting and drawing techniques, see Benjamin March, "Linear Perspective in Chinese Painting," *Eastern Art* 3 (1931): 113–39; George Rowley, *Principles of Chinese Painting* (Princeton: Princeton University Press, 1959).
[75] Ehrenberg, *Mapping the World*, 18. See also Amir D. Aczel, *The Riddle of the Compass: The Invention that Changed the World* (New York: Houghton Mifflin Harcourt, 2001).

mapmaking itself.[76] Ptolemy's developments in cartographical thinking were not rediscovered by the Western world until the fifteenth century. Christian philosophers, after the collapse of the Roman Empire, had the dominant influence over pictorial understandings of the world and the relapse to a flat interpretation of the earth.[77]

In the Islamic world, however, Ptolemy's work was translated into Arabic around the ninth century, which produced two traditions of mapmakers. The first tradition emerged in the tenth century and represented by the school of al-Balkhi, which focused on Islamic religious practices and the Islamic world.[78] The second was represented by the school of al-Idrisi, who was influenced particularly by Ptolemy's treatise and ancient Greek cartographical traditions. Al-Idrisi also utilized Persian and Indian geographical ideas in his interpretations. At the same time, these stylized views of the world acknowledged Islamic practices and sociocultural values of Islamic society.[79] Although mapping sacred directions was commonplace in Asia, Europe, and Middle East traditions of mapmaking, only Islamic maps presented a consistent object imbued with religious significance. Islamic maps honored the direction and location of the holy Ka'ba, the religious shrine in the great mosque of Mecca.[80]

In Europe, several styles of drawing, as well as ways of conceiving and representing space emerged and were practiced up until the end of the fifteenth century. They included way-finding road maps drawn around 335 AD by the Romans, medieval Christian representations indicated in the tripartite *mappaemundi*, and portolan charts of the thirteenth century, which were

---

[76] Ehrenberg, *Mapping the World*, 9. See also Lieberman, *Integration on the Mainland.*

[77] Ehrenberg, *Mapping the World*, 10. See also Tony Campbell and Marcel Destombes, *The Earliest Printed Maps 1472–1500*, ed. International Geographical Union Working Group on Early Maps (London: British Library, 1987).

[78] Ehrenberg, *Mapping the World*, 35. See also S. Maqbul Ahmad, "Kharita," in *The Encyclopedia of Islam: New Edition*, eds. C. E. Bosworth et al. (Leiden: Brill Academic Publishers, 2005).

[79] Ehrenberg, *Mapping the World*, 10. See also John P. Snyder, *Flattening the Earth: Two Thousand Years of Map Projections* (Chicago: University of Chicago Press, 1993); Norman J. Thrower, *Maps and Civilisation: Cartography in Culture and Society.* (Chicago: University of Chicago Press, 1996); Peter Whitfield, *The Image of the World: 20 Centuries of World Maps.* (London: The British Library, 2010).

[80] Ehrenberg, *Mapping the World*, 39. See also David A. King, *World Maps for Finding the Direction and Distance to Mecca: Innovation and Tradition in Islamic Science*, eds. H. Daiber and D. Pingree, vol. xxxvi of *Islamic Philosophy Theology and Science* (Leiden, The Netherlands: Brill, 1999).

essentially used for seafaring purposes. In addition, there was the crossover of these traditions in Cresques Abraham's map of 1375, in which religious and geographical information coexisted.[81]

Ptolemy's geographical works were circulated more widely by the fifteenth century, following the invention of the printing press, and the Western world again looked to interpreting the world through a new mode of scientific cartography. By the mid to late sixteenth century, the cartographical center of Europe had shifted from Italy, to Belgium, and the Netherlands, and Ptolemaic principles, which were subsequently abandoned for the Mercatorial worldview.[82] Gerardus Mercator's map of 1569 was published in Abraham Ortelius's atlas of 1570 as well as portrayed in Edward Wright's 1599 engraving. As a tool for navigation, Mercator's projection served to facilitate the ease of determining the navigator's position on one compass setting. From a geographical perspective, the projection presents a widely distorted view of the world: Africa, although roughly fourteen times larger than Greenland, appears to be a similar size with Mercator's view, and this misconception was subsequently conveyed to generations of school children.[83]

The commonalities between Asian, Middle Eastern, and European traditions of the premodern period include the often abstract nature of these maps—not adhering to an exact scale, and the often deliberate distorted or symbolic representation of geographic features. In some examples, certain features of the landscape might be highlighted with color and text, as was the case with portolan charts, while in other examples, a variety of viewing perspectives was adopted to convey a scene.[84] Roads, lakes, and boundaries might be depicted in two dimensions, while mountains and buildings often adopted a three-dimensional projection. The level of abstraction of mappings transcended many cultural groups.

[81] Ehrenberg, *Mapping the World*, 16.
[82] Ibid., 11. See Whitfield, "Image of the World," for discussions and large graphic images of the many configurations of these mapping styles of the sixteenth century.
[83] Ehrenberg, *Mapping the World*, 111.
[84] See also, on portolan charts, Derek Howse and Michael Sanderson, *Sea Chart* (Newton Abbot, UK: David & Charles PLC, 1973).

1.2 Maps and Mapping Practices

In many premodern societies, both European and non-European, maps were produced for different purposes and through different techniques. They were also drawn in a variety of forms and employed for an equally varied number of uses. The activities and records of early explorers and traders, European, Arab, Chinese, or Indian, reveal that maps of their regions had existed, but were often constructed in impermanent media. Travelers and pilgrims often traveled without the use of a map by our conventional standard, but using either their experience or the experiences of others through narrative geographies and verbal accounts of places as way-finding mechanisms.[85] Verbal accounts extended ideas about a place beyond the notions of mere geography. In addition, the traversed space of pilgrimage offered an alternative spatial conception that was consistent with embodied religious and cosmographic concepts and symbolism. In pilgrimage, symbols, statues, and temples, together with myth and ritual practices, provided the dimensions of physical space.[86] Methods for conveying these ideas included textual mappings as outlined below.

Christian Jacob, in his essay "Mapping in the Mind: The Earth from Ancient Alexandria," studies ancient Greek texts and demonstrates that records of the role of small-scale maps existed in this society. According to Jacob, "the result of such research [establishing the precise limits of maps, from a visual, intellectual, and pragmatic perspective], in ancient Greece, clearly suggests the very limited functions and diffusions of maps in this society."[87] Utilizations of maps in ancient Greece were not necessarily related to pragmatics, such as finding one's way, because often the format and kinds of representations of the period were not suited to such technical uses. Discourse, narrative, text, and human memory supplanted the role of the map in constructing geographies for purposes such as travel, trade, colonial encounters, and exploration. Jacob states that:

---

[85] Ehrenberg, *Mapping the World*, 31. See also Cordell D. K. Yee, *Space and Place: Map-making East and West: Four Hundred Years of Western and Chinese Cartography* (Annapolis, MD: Saint John's College Press, 1996).

[86] Winichakul, *Siam Mapped*, 29.

[87] Christian Jacob, "Mapping in the Mind: The Earth from Ancient Alexandria," in Cosgrove, *Mappings*, 26.

> one should be aware that categories such as vision, audition, memory and imagination do not escape the variations of culture and of history, and that in a precise cultural surrounding, listening to geographical description could create for the listener mental forms as vivid and efficient as a world map.[88]

Jacob bases his assessment of the role of maps in early Greece on first-millennium authors, such as Strabo, and his volumes named "Geography," in which no graphic images of maps prevail throughout—only textual descriptions of earlier treatises and maps.[89] Strabo is important because he notes details of the mapping process and the organization of the collated knowledge, and divides it into, first, the travelers' perspective and their recount of their journey or itinerary, which Strabo referred to as "the conception of *akoé* (hearsay)."[90] Another aspect to hearsay, which Strabo also acknowledged, was that the transfer of information was typically relayed in "discourse" about the event, which added another dimension to the conceptualizing of the place experienced by the traveler and then imaginatively reconstructed by the geographer. The second major act of the geographer revolves around collated information being processed into both an empirical and a mental diagramming of the world.

The first Greek maps were not created out of a desire for a precise geography but rather to deliver some opinions about a larger intellectual enterprise. Answers to some of the following questions may have been entertained in ancient mapping: What was the extent of the cosmos? How might we account for genesis? What was the nature of the world? How did humans arrive on earth? Anaximander of Miletus, a sixth-century BC scholar, wrote his view of the world from a georeligious perspective, as opposed to the conventional view of the time of "mytho-poetical tales of Homer and Hesiod."[91]

---

[88] Ibid.
[89] Ibid., 27.
[90] Ibid., 33. See also Rudolph Arnheim, *Visual Thinking* (California: University of California Press, 1969) for other conceptions of cognitive mapping.
[91] Jacob, "Mapping in the Mind," 27. See also, for a general overview of sources for these genres of cartography, O. A. W. Dilke, *Greek and Roman Maps* (London: Johns Hopkins University Press, 1998).

Anaximander presented an interesting and holistic notion of the world, formed from geometrical metaphors of astronomical observation and calculation to explain natural phenomena. Accordingly, premodern Greek geography is a collaboration of knowledge based on mythical lands and tribes. It is also understood within a cosmological frame dependent on the sun's passage across the sky for the different seasons, together with observations of the natural world, in addition to incorporating aspects of myth, language, and a cultural narrative from poetry.[92] Even this early type of mapping provided tools for scholars to conceptualize the world in alternative ways to that of Homer. Despite the detail associated with thinking about space, being predominantly literary, such maps presented a format to order knowledge about places independently from discussions of itineraries. Jacob also claims that mapmakers were typically philosophers and scientists. Individual maps were the result of one's own view of the world, free from stipulations of political or social controls.[93] Controls eventually originated from the cartographers as part of a discipline, in that as they accrued more information for their mapping practices from their predecessors and contemporaries, certain signs and rules became commonly practiced. Therefore, cartography became part of the mathematical sciences and removed from everyday life, whereas descriptive and imaginative geographies continued to be a part of the realm of historical studies and writers.

1.3 Imaginative Geographies

Geography has not always and everywhere been viewed as a static object. In premodern times, geography was not viewed purely as objective reality that can always be mapped in the same manner and with the same outcome, nor was the mapmaker viewed as a self-conscious observer independent of the mapped subject. Although it can be argued that geography is always constructed imaginatively, this was particularly so in the premodern period. The fact that people of different cultures mapped their world differently shows that there was something intrinsically imaginational about mapping, and that "imagined geographies" were in essence the object of all mapping practices.

---

[92] Jacob, "Mapping in the Mind," 28; C H. Kahn, *Anaximander and the Origins of the Greek Cosmology* (New York: Hackett, 1960).
[93] Jacob, "Mapping in the Mind," 29.

Being highly imaginative, the premodern map, unlike its modern counterpart, does not reveal the same meanings universally to its viewers, or as Denis Cosgrove suggests, the map is exposed to "inescapable contextual influences which shape the outcome of such transfers [of information]."[94]

Imaginative geography is not just the product of mapping, nor is it restricted to the premodern periods. The belief of British colonists arriving in Australia that it was an unpopulated land is an example of imagined geographies. Of course, this was not true because Aborigines had inhabited the Australian continent for centuries prior to the invasion of colonizers. Although the notion of "imagined geographies" can be discussed from many angles, as Edward Said's, Derek Gregory's, and Gayatri Chakravorty Spivak's pioneering works on cultural studies and postcolonialism indicate, it is the cultural and religious perspectives that are the focus in this study.[95] This is predominantly because of their role in premodern times, having the greatest effect on the ways in which geography was viewed, understood, and mapped.

Premodern perceptions of geography present vastly different understandings of space compared with ours in modern times because notions of the sacred and the profane were integral to coming to terms with one's place in the world. Premodern geography was understood primarily in terms of an overlap of its geopolitical and georeligious implications. Therefore, maps were not generally commissioned with solely political preoccupations but rather with divine and political agendas interwoven.[96] Two examples of imaginational and sacred spatial principles are elaborated below.

---

[94] Cosgrove, "Introduction: Mapping Meaning," 3.

[95] Derek Gregory, *Geographical Imaginations* (Cambridge, MA: Blackwell Publishers, 1994) and see Edward W. Said, *Orientalism* (New York: Vintage, 1979); Gayatri Chakravorty Spivak, *A Critique of Postcolonial Reason: Toward a History of the Vanishing Present* (Cambridge, MA: Harvard University Press, 1999).

[96] Often the king as the head of state was also the closest earthly representative of divinity. Therefore, decisions about mapped territory had a number of layers of meaning, and thus it follows that maps were open to many levels of interpretation and intended for a number of different audiences. See Nancy Shatzman Steinhardt, "Mapping the Chinese City: The Image and the Reality," in *Envisioning the City—Six Studies in Urban Cartography*, ed. David Buisseret (Chicago: University of Chicago Press, 1998).

Geopiety and Fada'il: Sacred Terrains

Religious conceptions of space dominated many premodern cultures, and combinations of spatial experiences, visions, and socioreligious imagery conveyed through recitals and readings of religious texts created imaginative mappings of the world.[97] Mapping practices based upon these descriptions formed a broader corpus of work than our modern understandings of mappings provide. In the premodern world, religion motivated a certain reading of spatiality and a means through which to appreciate that experience, whereas in modern society, sociocultural signs and practices play a greater role in aiding our spatial experience and realizations of our immediate territory.[98] One example of conceptual mapping from the premodern Arab-Islamic context is the one anchored in the religious concept of *fada'il*. Although the concept of *fada'il* predates Islam, "it is used mainly as an adjective in panegyric literature to denote the virtues and merits of certain texts, individuals, cities, monuments or times."[99]

Fada'il writings in the beginnings of the Islamic religion contained comments associated with the Prophet and his companions. Through the identification of certain landmarks of religious significance, *fada'il* bestows these built forms within the socioreligious landscape with "spiritual, cosmological and eschatological significance."[100] In turn, a type of mapping occurs, by which the *fada'il* in its narrative sets the scene and creates a particular spatial experience for faithful followers of the scriptures. Therefore, in its construction of a certain geography of divine significance, the concept of *fada'il* promotes an imagined spatiality and view of the world, authorized by God and openly conveyed for his believers.

Imaginative geographies were also present in ancient Chinese culture, through the royal desire to represent their kingdom in its ideal state. According to Nancy Shatzman Steinhardt, this idealism originates in Chinese rulers' commitment to following a passage in a classical text of late first millennium BC called the *Rituals of Zhou*, which "offers a prescription for an ideal state

---

97 Samer Akkach, "Religious Mapping and the Spatiality of Difference," in *Thresholds*, (Sacrosanct, 25, 2002) : 69.
98 Ibid., 71.
99 Ibid. See also Gerald R. Tibbetts, *A Study of the Arabic Texts Containing Material on South-East Asia* (London: E. J. Brill, 1979).
100 Ibid., 73.

capital."[101] This text appears to be the recommendations of the Zhou people through until the middle of the third millennium BC, which lists essential criteria for a royal capital. Among the technical principles listed, there were also other requirements, such as the skills of a geomancer to orientate development and a temple to house the "ruler's ancestors on the east, altars for sacrifices to soil and grain on the west."[102]

This may seem utilitarian enough except that excavated sections of parts of the walls of cities from the Zhou and other dynasties, such as the Han (206 BC to 220 AD), indicate that there is little similarity between the Rituals of the Zhou and the physical nature of Chinese royal cities.[103] In essence, the accolades of the premodern Chinese map lay not in the agency of its medium, the reputation of its creator, nor its technical or aesthetic qualities, but in "the regal system it symbolized in which there was always a place for a ruler to sit facing south in the center."[104] Therefore, the map was the ideal and imaginational space, whereas the physical city was something different.

Steinhardt goes on to compare the imaginative geographies of the ancient Roman cities with those of early Chinese dynasties. Specifically, Steinhardt considers the idealized nature of the city of the king of Zhou as well as a town founded by Romulus. Romulus represents a town of antiquity that was based on a highly conceptualized model.[105] These comparisons contribute to the concept of the imagined geography in their submission to the concerns of "myth, and ritual as opposed to economics or hygiene in founding of a city."[106] The Romans and the Chinese believed the role of the geomancer and the orientation of the city was crucial to its eternal fate, although the divining tools were unique to each culture. The Romans used a liver and a process called haruspex, whereas

---

[101] Steinhardt, "Mapping the Chinese City," 1–2.
[102] Ibid., 2.
[103] Ibid., 3. See also Liangyong Wu, "A Brief History of Ancient Chinese City Planning," *Urbs et Regio* 38 (1986).
[104] Steinhardt, "Mapping the Chinese City," 30. See also Mei-Ling Hsu, "The Han Maps and Early Chinese Cartography," *Annals of the Association of American Geographers* 68, no. 1 (1978): 9–17.
[105] Joseph Rykwert, *The Idea of a Town* (New Jersey: Princeton, 1976), 27. Rykwert explains that, unlike today, in antiquity there was not necessarily one meaning per one word, and similarly, with the construction of a town, it was ceremonial practices constituted by words and actions that gave sacred meaning to the place. The site was further celebrated with annual festivals, entrenching its meaning and significance in future generations.
[106] Steinhardt, "Mapping the Chinese City," 9.

the Chinese used the bone of a sheep, cow, or pig, or even a tortoise shell, depending on which animal was available, and the process was known as scapulimancy or plastromancy.[107] Mappings conveyed cultural preoccupations with adhering to ritual practice over the conveying of actual geographical space.

Georeligious belief systems were paramount in premodern Chinese and Roman cultures for determining one's place in the world. Yi-fu Tuan argues, through the concept of "geopiety," that the spiritual connection with place was inscribed in religious canons across many cultures.[108] In premodernity for the Chinese, geopiety, as a belief system, intrinsically linked to an individual's imaginative conception of space, dictated every aspect of earthly life.[109] The notion of piety covers relationships between people with the land, human beings and their god or gods, as well as human beings with nature. *Hsiao*, the Chinese term for piety, translates as "filial piety" or "filiality"; hence, the duty of care for one's parents and ancestors are a dominant aspect of pious Chinese.[110] The notion of *Hsiao*, from the Confucian text *Hsaio Ching* (ca. 350–200 BC), implies that there is duty of care from a "ruler to subject, parents to children, for without reciprocity there cannot be harmony."[111] In accordance with the ancient text, rulers served heaven by being pious to their fathers and served the earth by being pious to their mothers, which in turn related to all of the spirits being appropriately satisfied and equated to a harmonious existence.

In premodern Chinese tradition, there was a distinct opposition between heaven and earth. In the Chinese language, *t'ie*, heaven, is considered more of a counterpart although of marginally lower status in its relationship to *ti,* earth, and *t'u* (soil) is given lower status again. In both Chinese and European cultures, chthonian gods were superior to sky deities; therefore, created landscapes and architecture, from tombs to altars of worship and sacrifice, of this period reflected

---

[107] Ibid., 10.

[108] Yi-Fu Tuan, "Geopiety: A Theme in Man's Attachment to Nature and Place," in *Geographies of the Mind: Essays in Historical Geosophy*, eds. David Lowenthal and Martyn J. Bowden, 11–39 (New York: Oxford University Press, 1976).

[109] Ibid., 11–12. "'Geopiety' is a term borrowed from John K. Wright to stand for a complex of relations between man and nature. 'Geo' means earth; earth refers to the planet, the globe or its surface vis-à-vis heaven; it is also the soil and, by extension, land, country, and nation. 'Piety' means reverence and attachment to one's family and homeland, and to the gods who protect them."

[110] Ibid., 12.

[111] Ibid., 14.

such relationships.[112] Sociocultural relationships between mother and father also responded to this belief system; the mother was the nurturer (earth), and the father commanded honor, affection, and reverence (heaven). Sacrificial rites for both heaven and earth were only open to the emperor based upon his position in society, whereas princes of feudal states were allowed to make offerings within their own jurisdictions to earthly spirits.[113] Mappings conveyed the complexities of these spiritual relationships and often acted as illustrations of the required modes of religious practices to achieve fulfillment in earthly life.

Tuan suggests that, in modern Chinese culture, geopiety occurs because human beings possess a deep emotional bond with nature. Tuan further suggests that modern concepts have attempted to emulate past sociocultural practices and meanings associated with place, such as environmental activism, "giving back to nature," and are connected to "profound human experiences" that were predominantly played out in religious practices in ancient worlds.[114] In the absence of the entrenched socioreligious meanings and connections with place of premodernity, the modern concept looks to other forms and practices to achieve sociocultural links to space and reason for how it may be used. In addition, it is a distinctively modern conception in which there is awareness that human existence has a reciprocal arrangement with nature and therefore needs to be nurtured. It is a type of spiritual acknowledgment of the land associated with modern times rather than a direct religious association.[115]

By contrast, places in the landscape for ancient Greeks and Romans had different degrees of holiness, and were termed *loca sacra*, pertaining to an area set aside by the state for a deity, and *loca religiosa* equated to local shrines or burial grounds for certain families.[116] These sites had great religious and cultural significance for the community and were not to be encroached. Even travelers, armies, or citizens passing by these sites, for example, if they were rivers, made

---

[112] Ibid., 15.

[113] Ibid., 17. See also Joseph Needham, *Mathematics and the Sciences of the Heavens and the Earth*, vol. 3 of *Science and Civilisation in China* (Cambridge: Cambridge University Press, 1954).

[114] Tuan, "Geopiety," 13.

[115] This may be seen as an example of projective mapping, which is elaborated on in Chapter Six. It demonstrates a motivation for designing a particular environment for particular peoples in the absence of premodern connections to lived space.

[116] Tuan, "Geopiety," 17.

sacrifices to appease the spirits, and by so doing, participated in aspects of the imaginative geography of local peoples. Tuan states for the case of ancient Greece:

> Centaurs were originally spirits of mountain torrents ... Nymphs were almost everywhere: they dwelt in mountains, cool caves, groves and meadows and by springs … The ancient landscape was full of numina or local powers; the topographical ensemble itself had power and so did its parts.[117]

In effect, human beings of premodern China, Greece, and Rome were surrounded by divinity and the semi-divine in that their ancestors as well as their parents had spiritual implications for their lives.

Paradise and Biblical Mapping

Catherine Delano Smith, in her paper "Maps as Art and Science," presents a study of biblical maps of the sixteenth century, highlighting the split between genres of biblical and geographical information in mapped European representations. Delano Smith uses the example of Peter Plancius's 1590 Dutch folio Bible, whereby

> the historical (biblical) content was now divorced from the geographical. Instead of showing each event in its proper geographical place on the map (indeed, *as* the map) the biblical scenes are encased within vignettes and set at the edge of the sheet. The geographical outlines are left in the center "pure" and uncluttered by historical narrative.[118]

---

[117] Tuan, "Geopiety," 18.

[118] Catherine Delano Smith, "Maps as Art and Science: Maps in Sixteenth Century Bibles," *Imago Mundi* 42 (1990): 73. See also Kenneth Nebenzahl, *Maps of the Bible Lands* (London: Times Books, 1987), plate 37, 100–1, whereby the number of vignettes are adjusted from two to fifteen in representing the Garden of Eden.

Although this act of splitting religious imagery from geography may appear insignificant, it meant that future Bible maps into the seventeenth century utilized this imagery as mere decoration, therefore trivializing the socioreligious message of the map to the periphery. Another point upon which Delano Smith dwells is the issue of meaning in the biblical maps, as the extent to which imaginative geography was permitted, and thus the extent to which its construction became a problematic issue for theologians of the Christian church. Much of the writings associated with any visual biblical representations were supposed to be read aloud and absorbed in addition to the images on the page. Therefore, the orator would construct a particular frame within which to direct the message, this in turn would be reinforced by the images and be open to the reader's interpretation. "In the case of bible-reading in the sixteenth century," Delano Smith explains, "either for Catholics or for Protestants, such freedom was dangerous; it would (and did) lead to nonconformity."[119] Prior to the Reformation, it was more customary for priests to read a select group of passages from the Bible and then offer their comments and explanations of the word of the Lord. Similarly, Bibles were not readily available to the masses until the late sixteenth century, which was also paralleled, with a degree of "standardization of religious thought [which] operated not only through the approved verbal text but also through the visual."[120] Religious symbolism was also well established in the Christian traditions by the sixteenth century; therefore, often the invisible conveyed the message more convincingly than what appeared for the map reader. The use of allegory was already well understood by Judeo-Christians. The "exegetic structure" depicted between Old and New Testaments were recognizable with the use of historical vignettes surrounding the Bible maps.[121]

---

[119] Delano Smith, "Maps as Art and Science," 78.

[120] Ibid. Mappings were used to appropriate religious thought. The world was categorized and identified in controlled ways by the religious hierarchy, and thus representations portrayed controlled interpretations of religious rhetoric. See also N. Goodman, *Languages of Art: An Approach to a Theory of Symbols* (New York: Bobbs-Merril, 1968), 31–32. It is suggested by Goodman that representation is proposed as an interpretation rather than an imitation of objects. In my study, this thinking underlies the difference and many variations between mapping and planning processes.

[121] Delano Smith, "Maps as Art and Science," 79. See also Sandra Hindman, *Text and Image in Fifteenth Century Illustrated Bibles* (Leiden: E. J. Brill, 1977), 28ff.

Literal interpretations of Bible maps were also important to the Protestant religion in the publication of Dutch texts in the fifteenth century that used geography to present each biblical event where and when it officially occurred.[122] Therefore, Bible maps were utilized as profound instruments in convincing the reader of both historical and geographical facts. Martin Luther was a key figure in promoting the individual's experience of scripture. To aid this process, Luther selected appropriate visual imagery and thematics, typically replicated in the form of Bible maps for his German-language Bibles. Therefore, for sixteenth-century European Christians, the georeligious experience was a carefully crafted production by the mapmaker and religious instructor to achieve the paradoxical–conformist imaginative geographic space.[123]

Locating Paradise

Alessandro Scafi, in his essay "Mapping Eden: Cartographies of Earthly Paradise," elaborates on medieval European views of the world loaded with theological discourse in understandings of notions of space and time. Medieval maps, based on views presented by Western Christianity, represented the Garden of Eden as a worldly event among other everyday happenings. Scafi states that:

> a map which shows Paradise as part of the world implicitly acknowledges, by the leap of imagination involved, the limits of human action … It is at once a confession of the transcendent craft of God operating on terrestrial space, and of the limits of human reasoning.[124]

Scafi successively negotiates the complex questions associated with the narrative of the Bible in analyzing scholarly and popularly held beliefs about

---

[122] Delano Smith, "Maps as Art and Science," 79. See also C. C. Christensen, "Luther's theology and the Uses of Religious Art," *The Lutheran Quarterly* 22 (1970): 147–65.

[123] Homann lead the revival of German cartography in the early eighteenth century. The vignettes were perhaps intended to instruct a spiritual connection with geography. Ehrenberg, 2006 *Mapping the World*, 130-31.

[124] Alessandro Scafi, "Mapping Eden: Cartographies of the Earthly Paradise," in Cosgrove, *Mappings*, 51.

the thinking and nature of the Garden of Eden. For example, was it a "real" or imagined place? Did it bridge the dichotomy of heaven and earth and exist somewhere between these theological places? In addition, however, the graphic translation of these theological dilemmas about the location of Eden are reproduced on a map, using religious imagery and concepts that may not be as easily understood in text alone.[125] The paradoxical notion of juxtaposing Eden on a world map suggested a graphical type of evidence for what was presented in the writings of the scriptures, also giving greater credence of one's path to paradise. However, there were many adversaries to these conceptual graphical depictions of paradise, ranging from poets to biblical scholars.[126]

It was difficult for both cartographers and scholars of theology to grasp a reality lying beyond the limits of what they were able to perceive on earth. Yet, to a certain degree for the mapmakers, through this process of grappling with depicting a realm outside of their physical experience, this meant that what was previously imaginative geography, when mapped, became a possible reality. Scafi states, "depiction changes the phenomenon depicted. It can make real something previously unreal."[127] Through the cartographic placement of the Garden of Eden and its visibility on the map, there is a distance between this previously conceptual idea and, say, the city of Jerusalem. Simply by being drawn, it is no longer the figment of one's imagination.

Yet in the process of drawings, there is an element of artistic license in the representation of the Garden of Eden. It does not share the same geographical merits of experienced space. Theologians have not been able to ascertain the dimensions or details of this sacred landscape even from within the textual descriptions of scripture. There are a few known elements contained within paradise, such as the tree of knowledge and the tree of life, but a three-

---

[125] Ibid., 53. See also A. Plaut, "Where is Paradise? The Mapping of a Myth," *The Map Collector* 29, no. 2 (1984): 2–7.

[126] Scafi, "Mapping Eden," 52. See also Daniel Dorling and David Fairbairn, *Mapping: Ways of Representing the World* (Harlow, UK: Prentice Hall, 1997).

[127] Scafi, "Mapping Eden," 53. See also G B. Caird, *The Language and Imagery of the Bible* (London: Duckworth, 1988). Where there are six different understandings proposed for the term "meaning," for the context of meaning of imagery in Bible scenes. J. B. Harley, "Meaning and Ambiguity in Tudor Cartography," in *English Map-Making 1500–1650*, ed. S. Tyacke, 22–45 (London: The British Library, 1983).

dimensional conception remains absent.[128] Therefore, indicating a place of no known spatial dimensions presents the first cartographic paradox. The second conundrum is that, by indicating the earthly paradise on a map, it is in effect bridging the spatial continuum believed to exist between earth and heaven. For theologians, the earthly paradise represents the boundary between eternity and time, so it is a zone where human time and space mix with notions of divine time and space, according to medieval orthodoxy.[129] Access to paradise is achievable for humans through their belief and worship of Christ, so there is an inference that a spiritual journey is possible into this geographical space of Eden as long as one does the right thing in the human world. Scafi states:

> Paradise is at once present and absent, and a point of transition. Situated in this world, it nevertheless belongs to a different dimension; its presence on maps is thus only a footprint, a physical expression of absence.[130]

Why were notions of imaginative geographies integral to the conceptual constructions of space in premodern cultures? One of the reasons behind this is the conceptual relationship between cartography and cosmography that blurs the boundary between the visible and the invisible, the concrete and the imaginary. These relationships changed radically in modern times, and consequently, the nature of the modern map and its role in society has been irreversibly changed.

1.4 Cosmography and Cartography

Unlike modern cartography, earthly and heavenly geographies were represented in ways that did not clearly differentiate between these spaces. Biblical paradise is a case in point. Cosmographic ideas as well as ways of seeing and understanding earthly existence were integrally related to holistic conceptions of the world. Harley and Woodward's series of volumes dedicated to the study of

---

[128] Scafi, "Mapping Eden," 55.
[129] Ibid., 57. See also R. Scribner, *For the Sake of the Simple Folk. Popular Propaganda for the German Reformation* (Cambridge: Cambridge University Press, 1981), 186.
[130] Scafi, "Mapping Eden," 59.

the history of cartography presents comprehensive insights into many premodern societies and their mapping practices. In addition, it highlights the inherent connections between cosmography, cosmology, and cartography, and shows the ways in which prevailing religious beliefs influenced the nature and production of maps in premodern times.

The difference between cosmography and cartography requires some clarifications. Generally speaking, cartography refers to the visual representation of earthly terrains as well as to the conceptualization and production of such visual representations in the form of maps. By contrast, cosmography is concerned with the visual representation of the cosmos as conceived and described through the science of cosmology. Thus, cosmography both describes and maps the known cosmos, including heavens and earth (and their equivalents in non-Western cultures) as well as other planets and cosmic spaces. By comparison, cosmology deals with the philosophical and scientific aspects of the universe: its order, structure, and laws with a specific focus on space and time.

Although cosmographies differ from culture to culture, there were some shared grounds in direction and alignment. The division of the circle into four parts, according to B. L. Gordon, was said to be the result of astronomical observations of prehistoric man that the North Star and the midday sun establish the north–south coordinates.[131] This hypothesis was later dismissed based on the irregularities associated with directions and latitudinal variations that occur between seasons. The alternative hypothesis is drawn from the relations of the human body, its symmetry, and the possibility of dividing it into four parts—front, back, left, and right. The concept of azimuth was then extrapolated with one of the four points located on the horizon, and perhaps another point fixed on an astronomical object.[132] In many cultures, the four cardinal points seem to be associated with more than just geographic position and location.[133]

---

[131] B. L. Gordon, "Sacred Directions, Orientation, and the Top of the Map," *History of Religions* 10, No. 3 (1971): 211. See also T. D. Atkinson, "Points of the Compass," in *Encyclopedia of Religion and Ethics* (New York, Charles Scribner & Son, 1955), 88.

[132] Gordon, "Sacred Directions," 212.

[133] For a detailed study of this point in the Islamic tradition, see Samer Akkach, *Cosmology and Architecture in Premodern Islam: An Architectural Reading of Mystical Ideas*, ed. Seyyed Hosein Nasr, SUNY Series on Islam (Albany, NY: State University of New York Press, 2005).

European traditions also gave precedence to direction in their cosmographical understandings. Directional importance in the Jewish Old Testament was illustrated with the division of the horizon into four parts, and this division appeared in the mythical landscapes that were imagined beyond earth. As an example, "four winds blow in heaven—supposed by the ancients to be located in the stars of Pleiades—and four rivers flow through paradise."[134]

East was considered the sacred direction for Jews because the holy City of Jerusalem was based upon an idealized map, in which directions had positive and negative connotations. By comparison, north had "evil" associations for premodern Jews:

> the spirit lifted me up between the earth and the heaven, and brought me in the vision of God to Jerusalem, to the door of the inner gate that looketh toward the north; where was the seat of the image of jealousy, which provoketh to jealousy.[135]

Most synagogues today are aligned from west to east, and the ark is situated to enable worshippers to face Jerusalem. Early Christians revered the east, and worshipping the rising sun was part of typical practice. By contrast, the west had satanic associations with the setting sun and assumptions about evil events or practices occurring after dark. Yet by the fourth century and following the victory of Constantine in Rome, Christians were no longer permitted to revere the sun, nor even look at it. This view was not consistent with the creation of the interior spaces of church architecture in the fifth century; typically, the choir assumed a location to the east. The interior of the church was divided into four parts, reflecting the four cardinal directions, and metaphorically recalling the universe. The altar, located to the east was said to be paradise. By the eighth century, the configuration of interior space had changed to permit the congregation to face east while addressing the altar. How did these religious alignments and beliefs resonate with and relate to cartography? Roman and Greek cosmographers started drawing their cosmographical plans to relate to their belief

---

[134] Gordon, "Sacred Directions," 212.
[135] Ibid., 212–13.

of earth being a flat disk of land with a surrounding sea. Similarly, to the orientation of church architecture, maps were drawn with the sacred easterly direction to the top of any world maps. The Madaba Mosaic from around 550 AD is the earliest known example of Byzantine cartography with east at the top.[136]

Larissa Bonfante, in "Etruscan Boundaries and Prophecy," highlights the connection between cosmography and religious beliefs of premodern Etruscan peoples through notions of boundaries. Boundaries possessed sacred significance in addition to their metaphorical and abstract qualities for the Etruscans.[137] The Etruscans were located in central Italy ca. 800–500 BCE, and considered themselves to be sophisticated and wealthy urban dwellers.[138] Archaeological evidence of this European period includes remains and indications of the lives of the Etruscans, which were characterized by extravagant private houses and lavish burial tombs. The Greeks were present in this region after 800 BCE and brought their alphabet, Greek mythology, and their monumental architecture.[139]

Yet the Etruscans held onto their religious traditions and rituals, as well as their language. The Romans passed on technical aspects of city building to the Etruscans such as surveying and measuring land areas and forming physical boundaries between properties, whereas the Etruscans were appreciated by the Romans for their craftsmanship and aptitude for building. Specific Etruscan rituals concerning the positioning and orientation of building through practices such as divination, which aided the appeasement of the gods, were also acknowledged by the Romans. The notion of boundaries that were traversed during these practices included cosmic, earthly, and the "*saecula* or boundaries of time," all of which were vitally important to Etruscans' understandings of their worldly existence.[140]

Etruscans believed the sky god presided over all earthly boundaries. According to Etruscan philosophy, the sky gods ordered that boundary stones be erected on corners of allotments because men could not be trusted to abide

---

[136] Gordon, "Sacred Directions," 217.

[137] Larissa Bonfante, "Etruscan Boundaries and Prophecy," in Atkin and Rykwert, *Structure and Meaning*, 153.

[138] BCE. Refers to Before the Common Era.

[139] Bonfante, "Etruscan Boundaries and Prophecy," 153. See also T. J. Cornell, *The Beginnings of Rome* (London: Routledge, 1995).

[140] Bonfante, "Etruscan Boundaries and Prophecy," 153–54.

by any form of identification of territory other than physically defined spaces. This prophecy was said to have dated back to ca. 90 BCE, and highlights a clear relationship between the role of the gods and men as well as the divine significance of the notion of boundary. Roman records identify the Etruscan prophecies associated with concepts of boundaries as also putting forward ideas about geographical area as distinct from religious notions of space. This geographical area gave peoples of the region, or those who had any kind of association with the land, an identity as being from a certain place. Previously, identity reflected being part of a religion or speaking a particular language.[141]

Therefore, the Romans attributed communications between gods and the people, on subjects such as spatial boundaries, to the talent of the Etruscans. According to the Roman author Seneca,

> the difference between us and the Etruscans is … that, while we believe lightning is released as result of a collision of clouds, they believe that clouds collide so as to cause lightning. For since they attribute everything to the gods' will, they believe, not that things have a meaning in so far as they occur, but rather that they happen because they *must* have a meaning.[142]

Other boundary demarcations important to Etruscans were those between the living and the dead. Stone markers usually identified a grave site or family tomb, as well as the sex of the inhabitants, or whether it was an individual, a couple, or family group. The demarcations were explicit in their representation of the sexual organs of the inhabitants.[143] Such demarcations also acted as the thresholds between earthly existence and an underworld existence. The representational form was significant for the afterlife experience, but they were also important boundary markers. Therefore, it follows

---

[141] Ibid., 155. See also W. V. Harris, *Rome in Etruria and Umbria* (Oxford: Oxford University Press, 1971).
[142] *Quaestiones Naturales* 32.2 in Bonfante, "Etruscan Boundaries and Prophecy," 155.
[143] Bonfante,"Etruscan Boundaries and Prophecy," 158.

that the transition between earthly existence and a presence in an afterlife would also be the focus of Etruscan philosophy.[144]

B. L. Gordon indicates that geographic symbolism becomes more complicated when considering Muslim practices.[145] Gordon suggests that directions are important for Muslims, which is recognized in the direction of prayer being toward the *qibla*, as well as in words such as "azimuth," "zenith," and "nadir," which came to the English language from Arabic.[146] The Prophet Mohammed had also prayed to the holy city of Jerusalem before his revelation, which prioritized Mecca over Jerusalem. Many reverent practices of this period are still current for today's Muslims. For example, any animal slaughtered for its meat must have its head turned toward Mecca; the *qibla* is always faced in prayer. Any practices such as spitting or toileting in the direction of the *qibla* are forbidden. East was considered the most sacred direction for Christians, but south predominated at the top of the map for medieval Muslims.[147] The determination of the sacred direction (*qibla*) toward the *Ka'ba* in Mecca was a problem for Muslim scientists and scholars of the sacred law of Islam (*fuqahā*) for centuries.[148] Folk astronomy, stemming from the seventh century, is now understood for determining the *qibla*. One theory has each sector of the world around the *Ka'ba* related with the segment of the perimeter of the edifice.[149]

Ahmet T. Karamustafa, in "Introduction to Islamic Maps," identifies the complex nature of the different cartographical traditions of premodern Islam. In his view, there were two main schools of thought from 700 to 1850 AD, stemming from either a theoretical perspective or an empirical type of cartography.[150] Greek philosophical influence was most important to Islamic

---

144 Ibid., 159. Plato in *Timaeus* describes humans as being created by the material of the cosmos, and it is implied they also have a soul. Cosgrove believes these ideas relate to the understanding that "the form and spirit of the cosmos are mapped into our material being." Therefore, by focusing on the role of humanity, the field of cosmography incorporates a geometric dimension. If earth is at the center, with humans as observers of movements of the cosmos and planetary circulation occurring around the earth, then earth acts as a theatre for human existence.

145 Gordon, "Sacred Directions," 214.

146 Ibid., 218.

147 Ibid.

148 King, *World Maps*, 47.

149 Ibid.

150 Ahmet T. Karamustafa, "Introduction to Islamic Maps," in *Cartography in the Traditional Islamic and South Asian Societies*, vol. 2, bk. 1, *History of Cartography*, eds. J. B. Harley and David Woodward, 3–11 (Chicago: University of Chicago Press, 1992), 3.

cartography by the end of the tenth century. The majority of Greek scientific works had been translated into Arabic and, according to Karamustafa, appropriations of these ideas within the Islamic scientific context had a profound effect on the emergent Islamic civilization.[151] In addition, the majority of premodern Islamic maps were part of larger manuscripts. Therefore, the relationship of text to understandings of the mapped information was paramount in the Islamic context and led Karamustafa to question to what extent in premodern Islamic traditions was there an independent cartographic discipline.[152] This reveals a hierarchical arranging of cartographic examples from a level at which the text and narrative dominate the visual imagery to a level at which the text informs but is not crucial to deciphering the story of the map. As was often the case in other premodern societies, interpretations, readings, and the study of cartography were solely the domain of the literati, highly educated scholars, and philosophers.

Karamustafa notes that there are clear gaps in the study of premodern Islamic maps in that there is seemingly little explanation as to why mathematical, celestial, and geographic knowledge evident to scholars of the time and denoted in historical texts were not represented in cartographic terms.[153] In addition, by comparison with European examples of the premodern period, there is a distinct difference in the absence of iconographic figures in Islamic maps. Yet Oya Pancaroğlu believes that medieval images were referenced frequently in Islamic texts on cosmography, and the "ultimate value of images … is a means to understand God."[154] However, Pancaroğlu suggests that the images were adopted to evoke emotional as well as intellectual responses in order for the reader to arrive at a life lesson.

Chinese cosmographic conceptions and notions of direction were at odds with the European traditions because the latter were concerned with east–west alignments.[155] By contrast, alignment of axes north–south was the

---

[151] Ibid., 4.
[152] Ibid., 5.
[153] Ibid., 7.
[154] O. Pancaroğlu, "Signs in the Horizon: Concepts of Image and Boundary in a Medieval Persian Cosmography," *Anthropology and Aesthetics, Islamic Arts* 43 (Spring 2003): 32.
[155] Gordon, "Sacred Directions," 218. See also Henderson, 1994, suggests the most controversial cosmographical model in classical China was the *Luo Shu* (Luo River Writing) in addition to *Hetu*

predominating approach for the Chinese.[156] Although both directions were valued, southward reflected the orientation of the gods and the logic behind the southward-pointing compass for the Chinese.[157] Therefore, worshippers would face north, and in burial, the deceased's head would be positioned to the north. It was not until the thirteenth century, when a Mongol Emperor and conqueror, who was influenced by Muslim mapmakers, stated maps should have south to the top, that this practice was adopted in China.[158] Up until the end of the twelfth century, maps of Chinese authorship were always indicated with north to the top.

Premodern China did not have the same preoccupations with the cosmos as the West. John B. Henderson, in his study, "Chinese Cosmographical Thought: The High Intellectual Tradition," states "that China compris[ing] 'all under heaven' (*tianxia*)" may have been a reason for not being concerned with cosmography and with representing the world as whole.[159] Traditional Chinese cosmography typically took the form of charts and represented more of a localized view of the world. Elements represented on the charts included agrarian concerns, architectural features, and urban dimensions of cities, but not the shape of the world or its imagined extents. However, the charts did presuppose connections between localized issues and the macrocosm of the world, although they were not often graphically represented. Textual accompaniment to Chinese cosmographical charts was also vital from the Song era (960–1279).[160] Graphic depictions were subordinate to narrative. The *Book of Changes* or *Yi jing* implies that perhaps earlier cosmographical works had a different emphasis between text and drawings, yet because there is not sufficient

(Yellow River chart). These were not applied directly to physical spatial models, but the *Luo Shu* was regarded as the original source of the magic square model and were also interpreted as the base to all *tu*, the term applied to all forms of Chinese graphic representations. Accordingly, the *Luo Shu* were at the core and the predominant source of what the ancient sage kings defined as the order of the world. Again, however, as there are only a few examples remaining of such concepts, and only from the post-Song era, their precise cosmographic form is widely debated.

[156] Gordon, "Sacred Directions," 218.

[157] Ibid.

[158] Ibid.

[159] John B. Henderson, "Chinese Cosmographical Thought: The High Intellectual Tradition," in Harley and Woodward, *Cartography in the Traditional East and Southeast Asian Societies*, 203.

[160] Ibid. See also Michael Sullivan, *The Arts of China*, 3rd ed. (Berkeley and Los Angeles: University of California Press, 1984).

evidence remaining to substantiate such implications, it has not been confirmed by scholars.[161] Because of the lack of physical artifacts from the pre-Song era, Henderson states that the majority of his study uses verbal accounts of Chinese cosmographical thoughts in the Han era (206 BC–220 AD).[162]

The conventional cosmographical model of the Han era was a square divided into nine equal smaller squares, presenting a three-by-three grid. This diagrammatic form was the base for many architectural, agricultural, political, astronomical, and urban concepts and plans. Henderson suggests that "this form was at least as important and ubiquitous in premodern Chinese cosmography as the circle was in Greek, medieval European, and Islamic cosmography."[163] The basic square divided in this way represented the basis to all order and proportion for Chinese culture during the Han period. The rectilinear geometries were also a feature of earlier dynasties than the Han because artifacts such as pottery display rectangles as decoration. Yet the square cosmographical charts with a clear systematic division of space did not appear until the Han era. The origins of the magic square system are not known by Sinologists, and possibilities include that an ancient sage king, Yu the Great, saw the pattern on the back of a turtle and thus adopted it from the natural world.[164] Most notably, however, the magic square was conceived for practical applications rather than just as an image of a cosmographical view of the world. It was envisaged that an arrangement of proportion and order would benefit human experience and society through architectural, medical, and agrarian practices. The applications in agriculture included the "well-field system," and for the architectural sector, there was the "nine-palace formation," the former being to ensure good harvests and the latter to maintain harmony in the imperial kingdom.[165] The well-field system

---

[161] Henderson, "Chinese Cosmographical Thought," 203.
[162] Ibid. Cammann Schuyler, "The Magic Square of Three in Old Chinese Philosophy and Religion," *History of Religions* 1 (Summer 1961): 37–79.
[163] Henderson, "Chinese Cosmographical Thought," 203. The magic square also formed a basis for geopolitical landscapes within China, dating from the Han era. A series of nine regions based on the eight cardinal directions and the center were documented in a first-century text, *Li ji* (Records of rituals), which states "within the four seas, there are nine regions, each a thousand *li* squared." Therefore, there was the central royal capital surrounded by concentric squares of regional areas, a structure that formed a complete geometrization of the classical urban Chinese landscape in the Han period.
[164] Ibid., 204.
[165] Ibid., 204–5.

was the basis for groups of eight families to have a portion of land each with the middle portion of land assigned as a shared allotment for the benefit of all. Throughout Chinese history, Confucian scholars have called for its reintroduction based on its being central to social rights because everything is measured and distributed equally. The well-field system also became a feature of Japanese agricultural arrangements.

There was also another conceptualization of cosmographical concepts to achieve what was known as the "field-allocation," or *fenye* system, which achieved more of a macrocosmic view. Heaven and earth were connected, and each of the nine regions related to a celestial field. By the Han era, the *fenye* system had developed into an astrological process whereby regions on earth became directly related to regions within the heavens. Initially, nine regions related to nine celestial fields, but later Han documentation suggests that there was an extension of astronomical data, with "twenty-eight lunar lodges, the twelve stations of the Jupiter cycle, and the nine paths of the moon."[166] The implications of these astronomical regions in relation to earthly regions meant that measurement of the boundaries of each was required. As in the other cosmological systems mentioned, *fenye* had practical applications. Any untoward changes in the cosmos, such as the appearance of comets, meant that there were direct consequences on earth. For example, the region relating to the celestial territory containing the comet might signal the end of political control by that leader.

Feng Shui: Cosmography and Cartography

An aspect of cosmography in ancient China was geomancy, or *feng shui* (literally, wind and water). Two schools of thought in geomancy developed in the Song era: the first was closely aligned with cosmological concepts such as *yin* and *yang* and *wuxing* (five phases), and the second was associated with the topographical nature of the landscape.[167] Chinese cosmographical influence extended most particularly to the realization of urban forms around the end of the Han era. Classical Beijing was built around cosmographical concepts, such as the

---

[166] Ibid., 209.

[167] Ibid., 219. See also Jeffrey F. Meyer, "*Feng-Shui* of the Chinese City," *History of Religions* 18 (November 1978): 138–55.

alignment of its rectilinear plan, its walls, and the gates within its walls with the points of the compass.[168] Adjustments to the ideal city plan were made based on geomancing requirements, accommodating the existing built structures, and allowing for numerous topographical limitations. Chinese urban models based on cosmography extended also to premodern Korea and Japan. The former school of thought depended on a geomantic compass, whose readings were utilized in relation to applicable measurements of space and time.

All of these readings and analyses, leading to a form of mapping and layout of site and buildings, were the specialization of a professional geomancer. The latter approach relied less on issues of the cosmos and more on topological characteristics of the terrain in question. Practitioners of this school made schematic topographical mappings of the site, which also indicated where the main energies or *qi* of the landscape lay. Metaphors were also drawn between features of the landscape and mythical characters, parts of the human body as well as celestial regions. Henderson states:

> Cosmologically, the *feng shui* spot is the meeting place of the heavenly *qi* and the earthly *qi*, where the *yang* and *yin*, or male and female influences, are most closely intertwined and most intensely concentrated.[169]

Geomancy and landscape painting may be related to the development of cartography in China. However, a conflict between principles of geomancy and pure geometric cosmography, in achieving a completely conformist classical imperial Chinese city, existed.[170]

1.5 Concluding Remarks

In summary, premodern mappings aligned sacred and profane worlds, and in many graphic interpretations of these spaces, there was little distinction between physical and spiritual worlds. Cosmographical and

[168] Henderson, "Chinese Cosmographical Thought," 211.
[169] Ibid., 217.
[170] Ibid. See also Richard J. Smith, *Fortune-Tellers and Philosophers: Divination in Traditional Chinese Society*, Westview Special Studies on China, (Boulder: Westview Press, 1991). Consider

cartographical representations were typically accompanied by text and narrative to explain trajectories of existence for peoples of premodern societies. These maps convey sets of relations and values pertaining to premodern peoples discernibly different from modern mappings.

CHAPTER TWO
Forms of Premodern Mapping

Premodern Southeast Asian cartographic and cosmographic practices were heavily influenced by their trading partners in the region, and to a lesser extent, traders from the Middle East and Europe. This chapter shows that the scope of different mapping genres that have surfaced in the Southeast Asian region was due to its rich and diverse cross-cultural interactions. The commonality derived from this broad spectrum of types seems to relate to issues of quality and meaning brought to the lives of peoples who witness or utilize the teachings of these often elaborate representations. As Schwartzberg notes, *Yuan shi* (*History of Yuan*), an official Chinese history compiled in 1369–70 by Song Lian and others, writes about a military campaign in Java in the thirteenth century. One of the leaders of the campaign, Raden Vijaya of the state of Kediri, in surrendering the battle, hands to the enemy a map and census record of the invaded country.[171] This is the first written notice known of the existence of a Southeast Asian map. Overall, the surviving corpus of premodern Southeast Asian maps is slim. The limited remaining collections do not necessarily equate to a lesser importance or degree of use of Southeast Asian mappings, but rather that climatic conditions, mapping materials and techniques, placed restrictions on their longevity.

From extant accounts by travelers, traders, and religious pilgrims, it is evident that Southeast Asian mappings provided a valuable sociocultural

---

[171] Joseph E. Schwartzberg, "Introduction to South Asian Cartography," in Harley and Woodward, *Cartography in the Traditional Islamic and South Asian Societies*, 697.

component to everyday life. Thomas Suárez, historian and antique map collector, divides Southeast Asian maps into four general categories:

> Those which are purely cosmographic (in this context meaning "metaphysical" or "spiritual") in nature, or otherwise non-geographic.
> Those which symbolically represent actual geographic features for religious or cosmographic purposes.
> Those which attempt to record true geography, whether by report or empirical observation.
> Itineraries, which might be written, memorized, or committed to song, that served to construct a mental image of time and space, direction and position, topography and landmarks.[172]

These categories, although not completely distinct, nor universal, demonstrate how maps and mapping processes were central to Southeast Asian life experiences, from the magical to the mundane.[173] The following discussions focus on some key examples of these maps to show the qualities they brought to quotidian existence for peoples of the region. Owing to the holistic nature of Southeast Asian methods of historical mapping practices, from the making to the realization of the map, rich sociocultural, religious, and geographical understanding and visualization are depicted.

2.1 Styles, Techniques, and Material in Mapping

Styles, techniques, and materials of Southeast Asian maps were highly variable. Because of the predominant influences of traders from India, China, Europe, and the Middle East, artisans, traditions, and religious practices also made their way to the region. Because of their preoccupation with imaginal, cosmological, and religious realities, premodern Southeast Asian maps were characterized by their lack of precision and accuracy. The key pragmatic traits that differentiated Southeast Asian maps from their European counterparts

---

[172] Suárez, Early Mapping, 33.
[173] Ibid. See also Kenneth Hall, *Maritime Trade and State Development in Early Southeast Asia* (Honolulu: University of Hawai'i Press, 1985).

included, first, that they were almost never drawn to a recognizable scale, second, that their projection differed, third, that they lacked the geographic grids of latitude and longitude, and fourth, that orientation varied by region. In addition, legends were not indicated, and dates of production were not stated.[174] The names of the mapmaker, surveyor, or artist were also typically omitted, and almost no maps demonstrated neatly constructed lines.[175]

Style of an Ideal City

Early Vietnamese mapmakers were influenced by ancient Chinese cartography, whereby mapmakers were employed to create a ruler's ideal for a city or kingdom. Therefore, graphic depictions presented to the king ranged from future expectations for his sovereignty to examples of sheer "mythology, fantasy, or forgery."[176] The mapmaker was typically a painter, skilled in both calligraphy and poetry. They were encouraged to maximize artistic license and to manufacture maps for purposes– be it military, mercantile, or glory.[177] In essence, throughout the mapping process, there was little emphasis placed on conveyance of a sense of truth or reality.

Peoples, from caravaners to pilgrims, traveled without conventional maps, instead utilizing their experiences, as a type of narrative geography, to negotiate landscapes.[178] Typically, from personal observations, conveyed through spoken words, songs, or poetry, information about places was often colorfully distributed. This method of translation extended ideas about a place beyond the notions of mere geography. Verbal mappings were one of the earliest mapping methods, which produced associations with place that were ideologically and imaginatively removed from our contemporary notion of a map.

---

174 Joseph E. Schwartzberg, "Conclusion to Southeast Asian Cartography," in Harley and Woodward, *Cartography in the Traditional East and Southeast Asian Societies*, 841. Burmese maps used grids corresponding to premodern Chinese models. "…conventional map signs such as for settlements, rivers, coastlines, and lakes [were] almost always drawn planimetrically, whereas others, especially for mountain and hill ranges, vegetation and prominent edifices (stupas, temples, monasteries, etc) [were] generally drawn in frontal perspective."
175 Ibid., 841–42.
176 Steinhardt, "Mapping the Chinese City," 9.
177 Ibid., 16.
178 Suárez, *Early Mapping*, 31.

Cosmological City Style

Premodern Asians had differing concepts of reality between cultural groups. For example, a shared understanding of critical concepts of boundary and territory did not exist. For Southeast Asians, the center-periphery notion of a kingdom was central to the understandings of territory, in contrast with European concepts, which considered territory as bounded space demarcated by borders. The *mandala* or "contained core," represented a sacred schematic of the cosmos in Indian philosophy.[179]

The transmigration of the soul, from an earthly existence to a continual cycle of rebirths, which was when transcendence occurred—a realm characterized by a vertically organized hierarchy of earth and series of heavens. This meant that souls were able to locate themselves within the universe, until they found their *nirvana* (Buddhism) or ultimate release from this cycle.[180] Cartographically, the visual representations of this process were typically in two dimensions but revealed a multidimensional universe, demonstrated expansively in the vertical plane. These maps displayed the passage of souls as well as the passage of the universe, which moved through similar phases. Time from a spatial dimension as well as temporal dimension addressed very large periods. For example, a Brahman was expected to live for over 300 hundred years; therefore, time and space had particular relevance for each ethnic group.

Natural features of each cosmos were often portrayed in mappings. Often, trees of outstanding proportions became synonymous with particular continents, and were located centrally in space, such as the *jambū* tree of India.[181] Shapes were also incorporated into cosmography to convey mountain ranges, which were represented by concentric rings or wedge-like forms depending on the view captured by the cosmography. The universe and earth were typically centered on an axis where Mount Meru, or Mount Sumeru (in the case of Southeast Asian examples), was at the center of existence. The characteristics of this sacred mountain varied dramatically between each of its graphic cultural interpretations,

---

[179] Hall, Maritime Trade, 9. See also Donald F. Lach and Edwin J. van Kley, A Century of Advance: Southeast Asia, vol. 3, bk 3, of Asia in the Making of Europe (Chicago: University of Chicago Press, 1993).

[180] Schwartzberg, "Introduction to South Asian Cartography," 334.

[181] Ibid., 335.

but in all instances, it was deemed physically too high for any human to ever attempt to reach its peak.[182] Unlike other cosmological mountains, Meru was conceived as having distinct layers where different supernatural existences may be accommodated.

Itinerary Mapping Styles

Another method used was the graphic itinerary, which explained routes for traversal between holy spaces and temples. The breadth of religious influences to the region came to Southeast Asia through many different avenues over time, and each set of religious teachings was never replaced in totality. Instead, the previous religious toutings were built upon to reflect new belief systems. The graphic translation of layered religious ideas was demonstrated with icons, statues, and textual mantras infiltrating representations of space.[183]

Techniques and Materials

Common processes of production in mapmaking included the use of paper, silk, wood, and stone to convey mapped information. In East Asia, silk was used as a map medium due to its elegance, resilience as a material, and weightlessness. However, it was very expensive to produce and was sought after by painters and calligraphers. Thus, it was abandoned in the mass production of Chinese maps.[184]

Stone was also a common medium because it was a permanent means of recording information. Stone mappings were usually undertaken in three stages: first, the drafting of the mapped information on another surface, such as paper or bark; second, the transfer of drafted information to the stone surface; and, third, the chiseling of the mapping into the surface as a permanent record. The latter stage involved two processes, the first being the outline chiseling followed by the final incision of the image and text. The stone image resulted in what is coined as "intaglio," producing a white-on-black image. Paper was utilized to make a

---

[182] Ibid.

[183] For example, even though there was a sense of the sacred and the profane, these notions were conceptually blurred and not endemic to one particular religion.

[184] Cordell D. K. Yee, "Chinese Cartography among the Arts: Objectivity, Subjectivity, Representation," in Harley and Woodward, *Cartography in the Traditional East and Southeast Asian Societies*, 137.

reproduction of the image by being pressed over with an ink-soaked cloth onto the stone-engraved surface. These techniques, along with reproducing images from woodblocks, stem from eighth-century image production in China. In contrast to the production of an image using stone, wood block carvings were known as "carved in relief," resulting in the image being black in outline on white paper.[185]

Each of these techniques served specific purposes in that paper reproductions were intended for temporary use, and the method of reading was to be horizontal. In contrast, stone carvings were intended for an intergenerational audience and were read vertically, so the reading, the message, and the technique of creation were geared to a different point of view. For the Chinese, the latter form of mappings recounted stories of cultural heritage, and such maps were important transmitters of information. The physical labor and dedication involved in creating such maps reflected how highly regarded and necessary the message was for future Chinese generations. Stone blocks were too weighty to transport, and rubbings of their maps was time consuming and more complex than reproduction by woodblocks.

Woodblocks enabled a greater level and ease of reproduction because of their portability. They were also relatively simple to alter by inserting wood plugs to reconstruct and amend the image. Wood was in good supply and therefore not expensive. The initial intention of a woodblock mapping was not to convey the detail and fine line work of their stone equivalents, nor did they have the compositional attention to detail in the quality of drawing or image layout. Often, woodblock mappings relied upon a textual backdrop to convey the full extent of the message.

The various techniques of mapping and painting in premodern East Asia have been considered historically comparable. Generally, however, map artifacts have been insufficient to draw a complete study of comparison. Since the discovery of some Han artifacts, the comparison between painting and mapping has been established through the notion of geography. Yee suggests, "it is perhaps

---

[185] Ibid. See also Li Xiaocong, *A Descriptive Catalogue of the Traditional Chinese Maps Collected in the Library of Congress*, ed. Library of Congress (Beijing: Wen Wu Chu Ban She, 2004).

connected etymologically to land configuration."[186] Since around the Song dynasty, landscape painting has been considered in China one of the most important genres of painting. Thus, it is interesting and significant to note that maps by renowned artists Zhang Heng and Pei Xiu feature in Zhang Yanyuan's catalog of paintings of the Tang period. There is one style of painting, namely, *jiehua*, which translates as ruled-line painting, that can be closely associated with the drafted technique of mapmaking. It is the only traditional style of Chinese painting that utilizes additional tools to the brush. For example, artists of this genre used calibrated rulers as well as drafter's squares, compasses, and other instruments, which aided accuracy in surveying and building measurements, such as a plumb line. This style of drawing seemed to be most common in the Song period, although using rulers and compasses for the purpose of recording types of representations to scale dates back to antiquity.[187] This technique did rely on producing images to scale, so it usually meant pictorial scale was prioritized over the natural scale of depicted objects. The intent of the representation was to alert the viewer to an emphasis of an object within the design of the picture, and if that were at odds with the natural dimensions of the object, it was typical practice to sacrifice the realistic dimensions. This attitude of Chinese mapmakers to prioritizing the pictorial nature of their compositions over the natural, continued into the Ming and Qing dynasties (1644–1911). Yee notes that there is a broad similarity between the way Chinese painters and mapmakers conceived of notions of space; however, such views were distinct from those of European cartographers of the period.

> The experience of space was dynamic and fluid, intimately related to one's experience of time. Space, emptiness, was regarded almost as an entity in itself—as such it was boundless and unlimited. Objects could be measured and defined; space could not be so fixed, since it changed with vantage point and time. As a result, no

[186] Yee, "Chinese Cartography," 139. See also Alexander C. Soper, "Early Chinese Landscape Painting," *Art Bulletin* 23 (1941): 141–64; Acker, *Some T'ang and Pre Tang Texts*.

[187] Yee, "Chinese Cartography," 142. See also Michael Sullivan, *Birth of Landscape Painting in China*, vol. 1 of *California studies in the History of Art* (London: Routledge and Kegan Paul, 1962).

> abstract geometrical system governed space, and points within it were not definable or delimitable in any absolute terms [whereas the European view of space included] a conception that defines space abstractly, as an entity that is bounded, static, and therefore organizable and measurable.[188]

During the Renaissance of Europe, when a diminishing ground plane was utilized to convey perspective in a static composition, the comparative dynamism of Chinese maps was achieved by multiple as well as variable viewpoints, in addition to several ground planes incorporated per object. This enabled the portrayal of a sequence of scenes within one scroll; in what we may understand as an action cartoon sequence, each image was revealed to the viewer, as the scroll was unraveled.[189]

According to an anthology of maps produced by the court of the Kangxi emperor (1654–1722), what defined a geographic map was an image that recorded a real place as opposed to an imaginary one. Therefore, a landscape painting assumes the artist greater artistic license than the mapmaker would otherwise possess. The degree to which each artist is permitted to stray from what is considered "reality" is not at all clear in such explanations of each craft.[190]In a geographic picture *tu* in Chinese, as well as the term *dili tu*, which refers to an additional definition of traditional Chinese cartographic practices, there is inferred the idea of "siting." This notion of siting, or *qi,* suggests a connection between traditional Chinese art and science, whereby "maps often depicted ideal siting configurations."[191] Therefore, the process for the Chinese mapmaker, which was recorded in an examination of a study of Qing hydrological works, as highlighted by the following quotation,

---

[188] Yee, "Chinese Cartography," 144–45. See also for the Chinese worldview based on organisms, Needham, *Mathematics and the Sciences of the Heavens and the Earth.*
[189] Ibid., 146. See also, for European influences on Chinese painting, James Cahill, *The Compelling Image: Nature and Style in Seventeenth-Century Chinese Painting* (Cambridge: Harvard University Press, 1982).
[190] Yee, "Chinese Cartography," 153.
[191] Ibid., 154. See also Steven J. Bennett, "Patterns of the Sky and Earth: A Chinese Science of Applied Cosmology," *Chinese Science* 3 (1978): 1–26.

> A map, like a painting, is not just a record, but is a product of the mapmakers' intuitive sense of underlying form—mapmaking involves abstraction of external details into something internal, a "mindscape." A cartographic image can thus represent not only physical appearances, but also the mapmaker's thoughts and reflections.[192]

Therefore, the map offered a connection between emotive experience and gaining knowledge about the world. According to Yee, in Chinese aesthetic theory, there is a fusion of the physical and psychological worlds. This overlay of the subjective and the objective was accentuated with maps being inscribed with text or poems. The appreciation of the viewer and the reader was on two levels, whereby the map engaged at a heightened emotive level when accompanied by text.[193]

2.2 Temporal and Permanent Mapping

Temporal mapping refers to the representation of space, which can be based on psychological and non-tangible associations, and may relate to changing conditions under various natural or supernatural influences. Temporal mapping extends to magical beliefs, relations between systems, and cosmological, astrological, or sociocultural practices that are open to different modes of interpretation. Temporal mapping, in the context of influences from China, included notions of ideal cities. Once these conceptions were realized on the ground, the urban form did not necessarily reflect original ideal intentions. Premodern itinerary mappings constructed sacred space for pilgrims through the linking of religious sites and temples; as such, they are another example of temporal, imaginative space, while including some permanent built structures in their representations. The transition from temporal to permanent can be seen in the mapping processes of land ownership and accrual through certain sociocultural customs.

---

[192] Yee, "Chinese Cartography," 158. See also Michael Sullivan, *The Three Perfections: Chinese Painting, Poetry and Calligraphy* (London: Thames and Hudson, 1974); Shen C. Y. Fu et al. *Traces of the Brush: Studies in Chinese Calligraphy* (New Haven: Yale University Press, 1977).
[193] Yee, "Chinese Cartography," 162–63.

Mapping the Astrological and Imaginal

One example of temporal mapping is the ancient Brahmans' conception of the cosmos as a "tortoise," with "its arched shell the heaven, its flat underside the earth."[194] The origin of this concept is dated to approximately mid-first millennium BC, but it evolved in the later *Purānas* around the sixth century AD. The concept of the tortoise featured in astrology and found its way into important works of South Asian astronomy.

Although data relating to this concept corresponded with actual geographical knowledge of its time, it did not appear to plausibly lead to an explanation of the ancient topography of India. Schwartzberg pragmatically suggests this is due to the dilemma of reconciling the actual geographical map of India by physically placing it into the shape of a tortoise, and subsequently making any sense of it.[195] Pictures of magical animals and objects as well as letters and numbers associated with astrology also guided practitioners of divination. Yet divination in India is still commonly practiced from the perspectives of all socioeconomic groups within society. It is usually accompanied by directions of almanacs "containing diagrams with terrestrial spatial referents."[196]

Overall, there appears to be no clear delineation between imaginative conceptions and reality in Indian cosmographies, although it was understood between Jainism, Buddhism, and Hinduism that earth was only a small part of more complex understandings of the greater universe. However, details and ideas about these remote other worlds were represented with a degree of vagueness.[197]

Jains' conceptions of their universe were far more complex than those of Hinduism, although they shared some similarities with vertical depictions of spaces, such as hells, heavens, earth, and netherworlds. Series of universes only filled a portion of cosmic space for the Jains. Void space within the universe was critical to the Jains because both matter and souls were deemed impenetrable. According to the Jains, our universe was composed of a series of netherworlds

---

194 Schwartzberg, "Introduction to South Asian Cartography," 337.
195 Ibid., 338.
196 Ibid., 339. See also D. C. Sircar, *Studies in the Geography of Ancient and Medieval India* (India: Motilal Banarsidass, 1971).
197 Gari Ledyard, "Cartography in Korea," in Harley and Woodward, *Cartography in the Traditional East and Southeast Asian Societies*, 341.

above and below the world of man. Each of the heavens and hells within the universe had its own peculiarities, which meant some were envisaged in the form of a figure of a man or woman.[198]

Mapping Miniatures: Temporal and the Natural

Vietnamese mappings predominantly adopted the style of Chinese cartographical traditions.[199] The Vietnamese word for "map" is *đồ*, meaning "illustration, drawing, plan and by extension map—*bản-đồ*."[200] However, the terminology closest to describing Vietnamese spatial representations is the term for country, *non-nước*, which equates to "mountain and water."[201] Mountain and water have symbolic qualities for the Vietnamese, in that they are used to convey temporal or supernatural stories as well as portray natural dimensions of landscape. Both magic and fertile stories are essential to Vietnamese conceptions of themselves in the world. Ancient Vietnamese temple complexes house miniature ponds and mountains, which are intended to act as scaled-down representational models of the universe. Tenth- and eleventh-century royal birthday rituals entailed mountains built out of bamboo, called "mountain of the South, *Nam Sơn*," being cast out on a boat in the middle of the river.[202] There were five peaks of which one was central (possibly Mount Meru), and among the four outer peaks was coiled a dragon. Similarities may be seen with cosmic ideas represented in Burmese and Cambodian temple complexes.

Mapping the Seen and the Unseen

The religious nature of premodern conceptions of space meant that understandings of the world extended from the seen to the unseen (refer Figure 1). The seen world was available to everybody, while the unseen was that in which one must believe and strive to learn about through the teachings of the Prophet

---

198 Ibid., 342.

199 John K. Whitmore, "Cartography in Vietnam," in Harley and Woodward, *Cartography in the Traditional East and Southeast Asian Societies*, 479.

200 Ibid.

201 Ibid. See also Keith Weller Taylor, *The Birth of Vietnam* (Berkeley and Los Angeles: University of California Press, 1983).

202 Whitmore, "Cartography in Vietnam," 479. See also D. R. SarDesai, *Southeast Asia Past & Present* (Boulder and San Francisco: Westview Press, 1994).

and abiding by the Quran. As Samer Akkach states for the case of Islamic teachings,

> The Quran stresses this polarity, describing God as "the Knower of the unseen and the seen" (13:9) and to him "belongs the unseen of the heavens and the earth" (16:77). The Quran repeatedly reminds the Muslims that no one knows the unseen except God: "And with him are the keys of the unseen. None but he knows them" (6:59).[203]

The unseen is the realm of the spirits and thus an imaginational space, whereas the seen is part of the natural world where understanding and perception is achieved through the senses. Analogy and metaphor are essential for teachings about the spiritual world, and key to connecting the divine with modes of human existence.[204] Mappings and cosmographies of the premodern Islamic world gave context to spiritual imaginings as well as connecting these ideas to the human existence on earth.

For example, Ibn Hawqal's world map from the eleventh century shows Ptolemy's earthly world as a circle surrounded by mountains, and the sea is enclosed beyond this circle. The map has the Ka'ba and Mecca at its center, which reflects the religious view of Islam at the time. South is oriented at the top of the map. Because Mecca and Medina were considered the center of the religious world, there were no lands indicated beyond this sphere.[205] Ibn Hawqal's world map is a highly stylized representation of predominantly the Islamic world, with major cities acknowledged with circles and mountains at the periphery in green, and rivers in blue except for the Nile whilst another major tributary is shown in red. This map demonstrates however, the shift from the temporal, unseen notions of spiritual lands of Mecca, to giving the town and space a physicality on the map.

---

[203] Akkach, *Cosmology and Architecture*, 29.
[204] Ibid.
[205] Ehrenberg, *Mapping the World*, 31.

Mapping Ownership

Mapping was not always considered a literal translation of a bounded space into a representation of that reality. In the following case, mapping is seen more as a process of establishing space through taking physical ownership over it.

In early Muslim towns, ownership of land played a key role in determining one's allocation of space. One had to carry out a series of strategies to establish ownership of land, particularly in the early Muslim garrison towns. The first stage was occupying the land; the second was passing it on to another prospective owner through a sale, or by gift of the land parcel. The final act of ownership was to pass it onto the next generation through inheritance.[206] The occupation of land in the establishment of early towns was the most common form of determining ownership because land was typically plentiful and, therefore, caused little dispute about who was the entitled landholder. However, as townships expanded, the system described above became the source of debate among Muslim jurists in the town. Land that was not claimed or used was designated and believed to be "dead" (*mawāt*) land. Such land parcels were defined in this way if no building or cultivation had occurred, no burials had been undertaken, or there were no common uses, such as pasture for the village animals or source of materials such as timber.[207] In due course, these lands could be revived by cultivating or building on them, and the people responsible would subsequently become the owners.

Demarcation (*ihtijār*) of land parcels or allotments was one method of reviving "dead" lands. This process referred to the "marking out" of a piece of land, either for a land parcel claimed by an individual or for land bestowed by a ruler, which effectively entitled a person to use the land rather than owning it. Those given land in this way were entitled to use it for three years. Arabic words associated with these practices often had numerous meanings and, therefore, historically have sometimes been interpreted out of their intended context. Jamel Akbar suggests "*khatt*" is one such word that has complicated and confused historians' understandings of this process of land ownership. *Khatt* literally translates as "line." Yet a fourteenth-century lexicographer, Ibn Manzur, stated

---

[206] Jamel Akbar, "Khatta and the Territorial Structures of Early Muslim Towns," (Dammam, Saudi Arabia: PhD thesis, King Faisal University), 22.
[207] Ibid., 22–23. See also Hakim, *Arabic-Islamic Cities*.

the definition of the noun *al-khattu* is equivalent to the "rectangular shape of a thing," such as the rectangular shape of an allotment of land, as well as "road."[208] Moreover, no direct translation into English exists for the verb "*khatta*." It can refer to marking out, outlining with the use of walls, straight lines and making rectangular things. In the majority of cases, *khatta* precedes the stage of building on the land. The term *khittah* suggests the notion of marking out a line with the intention of occupying a lot of ground, which in effect becomes the site, also known as the "pitch or holding." This definition combines ideas around the verb *khatta*, implying action, and the noun *khitta*, known as site development.[209] Understandings of the terms are important when differentiating between who made the decision to undertake the demarcations. For example, *khatta* implies the party decided for it, whereas *aqta'a* means someone else, usually a ruling authority, made the decision.

The *khitta* was typically expanded or increased depending on the size of the tribe being accommodated on the site. In other words, it was not used as a fixed planning unit but was responsive to the needs of its inhabitants. The central authority usually defined roads, and land between the major roads was not always organized into equal parts. Often, tribes would then subdivide their own *khitta* to suit the needs of the tribe. The only imposed external authority criteria was that dwellings should not exceed three stories for privacy between tribal neighbors.[210] Thus, within the *khitta* there were sub *khittas*, and these degrees of subdivision did not have to receive built space to their entirety. Open space was accommodated and ownership was not contested as a result. This was also a proviso for the expansion of the *khitta* to include relatives, who might move there from elsewhere. When land allocations did not permit expansion within the *khitta*, some inhabitants would move to join relatives in another location and newcomers would either be welcomed or follow their relatives to new lands. Shared space, which included cul-de-sacs, squares, streets and forecourts, were all collectively owned and operated within the *khitta*. Akbar suggests that the morphology of these early Muslim towns then came about by a multitude of decisions made on the micro scale by the inhabitants.

---

208 Akbar, "Khatta," 23.
209 Ibid; Nasr, *Ideals and Realities of Islam*.
210 Akbar, "Khatta," 29.

2.3 Textual and Figurative Mapping

Textual and figurative styles of mapping may be seen to overlap with the examples given under other themes of this chapter, and thus are indicative of the nature of the variety of influences and types within the Southeast Asian mapping genre. Textual and figurative geographies constructed space in a highly imaginative way; the former was reliant on written rather than drawn representations, and the latter indicated an imaginative element through its abstraction of figures on the map.

Textual Mapping

The textual recount of Zhang's rhapsody, cited below, contained phrases that may have been informed by maps, or in itself being a non-conventional depiction process: a verbal mapping of place.

> The first capital of the Han house
> Lay on the banks of the Wei River…
> To the left, there are…
> The double defiles of the Yao and the Han
> The barrier of the Taolin,
> Connected by the Two Hua peaks…
> To the right, there is…
> The gap of Longdi,
> Which partitions China from the barbarian lands…
> At its southern front, there are
> Zhongnan and Taiyi,
> Twisting upward tall and stately…
> At its northern rear, there are
> High hills and level plains…
> In the distance, there are
> Nine peaks and Sweet Springs.[211]

[211] *Wen xuan*, 4.13a–b; the translation is that of Knechtges, *Refined Literature*, 1:339 (note 13) in Yee, "Chinese Cartography," 132.

These descriptions of context placed the reader within the landscape and an imaginational construct similar to the way in which premodern maps tend to construct a perception of a place. The regular appearance in Han dynastic literature of the form of *fu* (rhapsody) was typically related to subjects of a geographic nature. A number of these literary sources also refer to aspects of surveying, forms of measurement, and maps.[212] A direct correlation existed for the Han Chinese, between the rule of the law as well as the notion of measurement, which was also the philosophical position of Han texts. "Law is the rule and measure of the empire, and the level and the plumb line of the ruler."[213] In effect, it suggests that a ruler may not be swayed or indicate any bias when passing the rule of the law, just like a suspended plumb line always remains true.

Political and literary uses of maps may be inferred from author Zhang Heng's rhapsody. In accordance with Zhang's descriptions of the creation of ancient Chinese cities, the emperor would refer to patterns, or *cai*, of other capitals when organizing the physical planning of his own city center. However, it is not clear whether those patterns were maps or verbal interpretations of places. There have been discoveries of maps, which have been dated prior to the record of Zhang's descriptions, uncovered in burial tombs.

Figurative Mapping

Urban life and layouts of urban centers feature in other types of Southeast Asian media also considered mappings. The predominant remaining corpus stems from Burmese origins, whereby the maps cover subjects such as neighboring centers and towns, among other topics.

One example is a military map of the Thai capital Ayutthaya from 1767, which was painted onto white *parabaik*, where a number of features of military purpose are identified (refer Figure 2).[214] The map was probably obtained for the purpose of invasion, which occurred in the year to which the map has been dated.

---

[212] Yee, "Chinese Cartography," 131.

[213] Huainanzi (ca. 120 BC) attributed to Liu An (d. 122 BC), in Huainanzi zhu (Commentary to Huainanzi, third century), ed. Gao You, chap. 9, see the modern edition (Taipei: Shijie Shuju, 1962), 140, all in Yee, "Chinese Cartography," 132.

[214] *Parabaik* is indigenous Burmese paper. Joseph E. Schwartzberg, "Southeast Asian Geographical Maps," in Harley and Woodward, *Cartography in the Traditional East and Southeast Asian Societies*, 790.

Notwithstanding cosmographies, this would be the oldest surviving Burmese map. The map contains little detail of the central and most grand compound, which was most likely the palace, but is quite elaborate in its examination of other features, such as temples, residences of high officials, and other administrative buildings.[215] The original map is elaborate in style and colorful in its depiction of housing, vegetation, and roads. At the same time, the figurative nature of the map has been exploited in the abundant use of wildlife, ranging from crocodiles to elephants. These figurative elements are exaggerated in scale, if the crocodile is compared with the size of the huts, and the vegetation is also highly stylized. To a certain degree, the stylization suggests an emphasis on features and aspects of the space portrayed on the map. For example, the map shows crocodiles entering the canal of the city, so presumably it is a space reserved for marine vessels and as a trading route, perhaps a source of sea or river food, rather than a swimming hole for the locals. In addition, the siting of rudimentary huts on stilts along the canal suggests certain humility of the population, typically, this style of housing in Southeast Asia is consistent with a fishing village.

2.4 Descriptive and Poetic Mapping

Narrative, descriptive, and poetic mapping traditions, from East Asian, South Asian, European, and Middle Eastern influences, were another means of prompting imaginational elements to be employed by the reader or observer within the landscape described. Literary and cosmographical route mappings offered conceptions of space familiar to local peoples because these ideas formed a part of everyday practices and teachings by community and spiritual leaders.

Literary Mapping

Through his poetry, a Chinese poet, Tao Qian (365–427), expressed his belief in realities conveyed by maps. This level of detail presented in maps of Chinese antiquity was highly convincing to its readers, and it is understandable that Tian Qian had such conviction about its representation of a reality. After reading the *Shanhai jing* (*The Classic of Mountains and Seas*), his first poetry series suggested that, after perusing some maps of mountains and seas, he felt that

---

[215] Ibid.,798.

he completely comprehended the universe. Tian Qian did not question the accuracy of the illustrations with which he was confronted. An extent of the material in the *Shanhai jing* was typically considered the subject of myth rather than reality. However, the contents of the present text of the *Shanhai jing* elaborated in detail about the hydrography, minerals, fauna, and flora pertaining to specific Chinese mountainous regions.[216]

Literary theorist Liu Xie (ca. 465–522) drew the connection between verbal, literary, and graphic depictions of landscape.

> Recently literature has been valued for verisimilitude [*xingsi*]. Writers perceive the true form of landscape, and penetrate the appearance of grass and plants ... Thus this technique of apt expression for form can be likened to the ink used for imprinting seals, for an impression made reproduces the seal to the smallest detail without further carving and cutting. Thus we can perceive appearances by looking at words, or know the season from the graphs.[217]

Liu Xie verifies the significance of language in conveying or representing reality, and proposes that language offers a "way of seeing."[218] The essence of reading the written word in Chinese languages is closely connected to the visual arts through the practice of calligraphy. Chinese characters are also visual representations of what is being conveyed through written and spoken language. Arts of Chinese poetry, calligraphy, and painting are known as the "three perfections," where the three disciplines are interconnected.[219] Learning the craft of Chinese painting required expertise in the art of calligraphy. Proportion and composition as well as the practical skill of coordinating the flow of the brush, were deemed essential attributes of a painter. Language and painting were considered equal in their power to represent both myth and reality, a connection

216 Yee, "Chinese Cartography," 133.
217 Liu, Wexin diaolong yizhu, chap. 46. (2:341–45) (note 23) in Yee, "Chinese Cartography," 133–34.
218 Yee, "Chinese Cartography," 134.
219 Ibid.

that evolved from the Han period when concepts of representation related to physical as well as the spiritual and cosmological world. This level of interconnectedness of the Chinese arts extended to the delivery of the physical representation. The notion of "breath-resonance," or *qiyun*, meant the "breath" was necessary in delivering, via the painted image, the innerness of the represented objects or composition, as well as the innerness of the artist.[220] The lack of "breath-resonance," according to some critics, restricted the quality of the work and the ability of the artist to achieve the best possible formal representation. Mapmakers also heeded such advice in their compositions.

Cosmographical Route Mapping

Burmese cosmographies, dated around the nineteenth century, proposed itineraries for pilgrims as well as informed the viewer of relations of cosmographical principles (refer Figure 3). For example, they typically depicted the location of twelve places on earth and their relation to Bodh Gaya (the place in northern India where the Buddha obtained enlightenment). The teachings of cosmographical mappings highlighted the blurring of existence between spatial relations both on earth and in the afterlife. For Southeast Asian peoples, how this transcendental path may be achieved, was of great importance.

Although this corpus of maps, geographically, is typically inaccurate, their layouts are by no means arbitrary and their textual accompaniments in the manuscripts indicate intent for their audience to make their way to these shrines in pilgrimage.[221] Route maps, non-conventional from European understandings, contributed to the spiritual and intellectual knowledge of Southeast Asian inhabitants. The detail of the discussions of the cosmographical manuscripts, according to Thai scholar Sommāi Premčhit, included instructions on how to construct stupas (*chedi*) in the style of the Mahābodhi Temple at Bodh Gaya in India.[222] The manuscripts contain diagrams, one of which identifies the holy

---

[220] Ibid., 135. See also James F. Cahill, "Confucian Elements in the Theory of Painting," in *The Confucian Persuasion*, ed. Arthur F. Wright (Stanford: Stanford University Press, 1960); Craig J. Reynolds, "Buddhist Cosmography in Thai History, with Special Reference to Nineteenth-Century Culture Change," *Journal of Asian Studies* 35 (1976): 203–20.
[221] Schwartzberg, "Southeast Asian Geographical Maps," 777.
[222] Ibid. See also Robert L. Brown, "Bodhgaya and South-east Asia," in *Bodhgaya: The Site of Enlightenment*, ed. Janice Leoshko (Bombay: Marg Publications, 1988).

places within proximity of Bodh Gaya. This particular diagram is known as a *yantra* (sacred visual charm), which was supposed to have originated from a powerful ancient monk who was reportedly from India around third century BC.

> The diagram consists of a square made up of nine smaller squares (three by three). Of the nine squares, the middle one represented as the "Buddha's throne" (at Bodh Gaya), is said to be the "center of all cities in the universe," while six others (all but the lower-center and lower-right squares, which show nothing but "sacred words") include the names, within triangles, of the principal sacred places in Champu Dipa ... associated with the Buddha in their (spatial) relationship to Bodh Gaya.[223]

No absolute hypothesis exists about the purpose of this map. Perhaps it is only a mythological itinerary because there is no consistency between travel distances and actuality. Sometimes the number of hours or days to reach a sacred site is accurate; alternatively, for other sites the time to achieve traversal is inaccurate, and there is no reasoning evident as to why or how this is so. There is also a suggestion that this was the itinerary of Buddha himself. These inconsistencies are not necessarily the problem of the mapmaker, but may be the lack of contextual knowledge in interpretation. The difficulties of interpreting such historical artifacts required a large amount of conjecture to establish meanings.

The map is known as "Small Portion of a very long Thai Route Map of an area between Nakhon Si Thammarat and Songkhla on the Malay Peninsula."[224] Notably, the town named in the title does not appear within the extents of the map. The map dates from the late seventeenth to the early eighteenth century, and it is part of a text referring to the Malay Peninsula, an area in the southern part of Thailand. This route map is oriented to the east and toward the Gulf of Thailand, and this section of the map was part of forty folios, of which there were possibly

---

[223] Schwartzberg, "Southeast Asian Geographical Maps," 777, 781. Acknowledging Thongchai Winichakul and David Wyatt of Cornell University's translation of Thai scholar Sommāi Premčhit's work.

[224] Schwartzberg, "Southeast Asian Geographical Maps," 784.

more. The route indicated is a combination of a land and a sea journey. One of the main intentions of the map was to indicate the rice fields, which relate to the temple Wat Phra Kho and those that belong to civilians. The dominant features of the map include temples, *chedis*, and other religious buildings. Also identified with ovals and squares are presumably a number of small and large settlements *en route*. Most of these lie in proximity of roads, yet sometimes what appear to be roads may also be streams or rivers with sporadic images of crocodiles and fish indicated. Vegetation is also indicated but, as is often the case with Burmese maps, according to Schwartzberg, the variety of vegetation is not necessarily representative in the figurative styles adopted by the mapmakers. Trees are usually indicated individually rather than as a forest and often are laden with fruit, with the odd appearance of a monkey or bird.[225]

In another route map titled here as "Shan map relating to a border dispute between (British Burma) and China along the Nam Mao River" (refer Figure 4) dated around the mid-nineteenth century, the text is in Chinese Shan, but there are other inscriptions on the map that are in Burmese.[226] The map is found in the Scott Collection, and it covers an area extending approximately fifteen miles, from a west-southwest direction to east-northeast direction, on either side of the Nam Mao River.[227] The yellow section and the northern part of the map, which is situated in China, reads, "all this is Möng Mäo territory," whereas the red section, which is located as part of Burma, reads, "the red is all Namkhan territory."[228] The influence of the British desire for identification of territorial control was expressed in the plethora of colors, which were also indicated on this map. Schwartzberg states that the map is relatively accurate in its conveying of geographical entities such as the riverscape because the modern version of the Nam Mao River and its path is very similar to the nineteenth-century map.

The only examples that Schwartzberg could source and that relate to the genre of geographical maps are some images that have been compiled on bamboo.[229] These are only some of the artifacts that have been obtained from

---

225 Ibid., 785.
226 Joseph E. Schwartzberg, "Cosmography in Southeast Asia," in Harley and Woodward, *Cartography in the Traditional East and Southeast Asian Societies*, plate 40.
227 Fifteen miles equates to approximately 24.1 kilometers.
228 Schwartzberg, "Southeast Asian Geographical Maps," 794.
229 Ibid., 795.

West Malaysian tribes known as the Negrito tribes. One of the tribal groups, called the Sakais, who survived on fishing, hunting, and gathering, partially supplemented by slash-and-burn subsistence, produced engravings on musical instruments or *tuang-tuang*.

Ceremonies were conducted using the magical powers of these instruments to achieve rainfall for their crops or protect them from evil spirits. Other practical applications required of these magical instruments were assisting in the location of housing materials. Schwartzberg describes the cartographic elements of these instruments as "highly abstract and far from obvious in meaning to an outsider."[230] However, their meaning was specific to the Sakais and aided their negotiation of elements within their everyday lived environment, for example, swamps, agricultural clearings, areas of a specific crop, locations of fish traps, anthills or areas consisting of deadly snakes, house sites and numerous other items and features integral to their existence in their local landscapes.[231] The variety of media utilized for recording mapped information, and the extents of specificities of sociocultural practices displayed have eluded many earlier cartographical studies.

2.5 Concluding Remarks

The descriptive and poetic nature of Southeast Asian mappings, addressing themes ranging from literary accounts to itineraries, suggests a sociocultural investment in the making and utilizing of a range of mappings to progress everyday life. These forms of mappings contributed to the religio-cultural conditions of Southeast Asian lived space, and in each type of mapping, there was an intrinsic purpose, which in turn gave meaning to the intangible and tangible qualities of human environments.

---

[230] Schwartzberg, "Southeast Asian Geographical Maps," 795. See also Vishvajit Pandya, "Movement and Space: Andamanese Cartography," *American Ethnologist* 17 (1990): 775–97.

[231] Schwartzberg, "Southeast Asian Geographical Maps," 795.

CHAPTER THREE
Mapping Settlements in Southeast Asia

Forms of settlement are embedded in our psyche: they influence our understandings and conceptions about dwelling as well as our social and cultural makeup. When we draw or write about settlement, it has to include something other than just settlement itself. Maps and illustrations speak graphically about specific qualities of the built environment, whether they are non-urban or urban. In this chapter, we examine the relationship between mapping and settlement, viewing them both as cultural practices. As shown in Chapter Two, premodern mappings contributed to the community by aiding the relay of information about belief systems, cultural practices, and expectations of the settled society. Geography was also an important aspect of premodern settlement, not only for pragmatic reasons related to natural topographical features, but also because of the sacred qualities associated with features, such as mountains in the natural landscape. Mappings of premodern geographies highlight these aspects of sacred space, which also affected settlement morphology. Some Southeast Asian representations of premodern space attempted to convey techniques, materials, and sacred space, as well as sociocultural rules and mythic practices to produce settlement. However, each site within the region and dominant ethnic group brought a diversity of values to the production of lived space, and representations of these spatial qualities attempted to convey these differences. As far as is discernible from the primary sources of this study, premodern Southeast Asian settlements were typically not planned or conceived in their entirety prior to their construction.

The notion of settlement may be understood here as a process that begins "through cognitive appropriation and modifications of our surroundings and the conveying of this information to others."[232] In terms of structure, it is "made up of explicit spatial and temporal relationships that define human existence."[233] As for settlement forms, they "provide culture with physical continuity and a setting for the individual within the community ... Like language, the shapes, forms, and relationships we gather from our environment and society create a way of thinking and apprehending the world."[234] In terms of producing human environments, this process may be seen as the rules of production of settlement, which are encoded in some way, and this codification of settlement practices may be played out through customs such as geomancing or the laws of indices, or through "analogic understandings of literary works."[235] These understandings go some way toward revealing the complex relationships that coexist between settlement forms and patterns, sociocultural practice, and landscape features that contribute to the milieu and idiosyncrasies of different societies.[236] Southeast Asian settlement sitings appear to have derived from these codes of understanding, with locations of built structures being specified because of the verbal instructions of the geomancer or royal ruler, conceived via ritual or festival, not as result of a mapped or illustrated representation.

3.1 Society, Culture, and Settlement

Mapping in premodern Southeast Asian society helps us to understand how citizens saw their role within the greater societal framework as well as their existence in earthly and non-earthly realms. These mappings also enable us to conceive of the importance of the relationship between physical and spiritual places in the world for Southeast Asian peoples. Because most of the surviving corpus of these representations do not predate the seventeenth century, and are not specifically urban mappings, other archaeological evidence (such as temple inscriptions) are utilized to confirm earlier motivations behind settlement

[232] Atkin and Rykwert, *Structure and Meaning*, ix.
[233] Ibid.
[234] Ibid. See also Stanislaus Fung, "The Language of Cultural Memory in Chinese Gardens," in Atkin and Rykwert, *Structure and Meaning*.
[235] Atkin and Rykwert, *Structure and Meaning*, 12.
[236] Ibid.

locations and formations in ancient Southeast Asian societies. Evers differentiates between the main urban settlement types in premodern Southeast Asian society, categorizing them as royal and sacred cities of inland states, commercial cities located as nodes in the trading network, or key smaller intermediate settlements that were both provincial capitals and involved in trade.[237] Different maps influenced different settlement types in a myriad of ways. In his archaeological studies, O'Connor successfully demonstrates that societal characteristics particular to Southeast Asian peoples have a long history in informing the complexities of settlement. These characteristics are based on established codes and rules of everyday practice, which vary between ethnic groups within communities. Under the following themes of sanctity, center, boundary, and geography, the discussion considers how these attributes and ordering principles contributed to forms of Southeast Asian settlement.

Sanctity

Sanctity was an important aspect of life for most premodern Southeast Asians, yet it affected the form of cities predominantly in the inland regions, which were often the base of ancient royal cities. These cities included Mandalay (Burma), Ayudhya (Siam), Angkor (Cambodia), Ava, and later Bangkok, which were characterized as strong capitals, propagating heavenly relations and practices, and having tectonic markers such as temples, palaces, and sacred space within their walls, with their settlement forms reflecting ancient Indian and Chinese urban concepts.[238] The main temple and/or palace represented the cosmological center of the capital of territorial control and administration, with the king coordinating the secular and sacred organization of the city.[239] Although there must have been aspects of planning behind the creation of these cities, it is not evident from the remaining corpus of maps that it was a completely preconceived process. Temple inscriptions and cosmologies convey the spiritual

---

[237] Hans-Dieter Evers and Rüdiger Korff, *Southeast Asian Urbanism: The Meaning and Power of Social Space* (Hamburg/New York: Lit Verlag/St. Martin's Press, 2000), 30.
[238] Ibid., 30–31.
[239] Ibid., 27. See also H. G. Quartich-Wales, *Siamese State Ceremonies: Their History and Function* (London: Bernard Quartich, 1931).

meanings and messages associated with settlement formation, but do not represent implicitly the settlement itself.

The following example describing the creation of Bangkok highlights ordering principles for realizing human settlement other than using the map. Instead, practices consisting of a series of rituals and the establishing of hierarchical relationships of elites within space contributed to the form and layout of this city.[240] Appeasement of the gods with ceremonial undertakings assisted the specific site, orientation, and key axial relationships of built form within the new city.

Bangkok, the capital of Siam, was planned via cosmological concepts, with the existing sacred temples integrated into the overall form of the city. The three main constructions that contributed to the city were the city pillar, the royal chapel, and the *Thung Phramen* or large public square. These three structures form a triangle of independent focal points within the city. To determine the location of the pillar, the gods had to be made aware of the proposal, and a ceremony that restaged the "founding of the world by Brahma" had to be undertaken to allocate a position for the guardian deity.[241] The royal temple represents the center of the country because it houses the most auspicious symbol of Buddhism, which is an expression of the King as the main supporter of Buddha.[242] The third component is *Thung Phramen*, which represents the world mountain, typically used for cremations as well as other public rituals. These three spaces and forms represent, in order, the guardian deity, which has power over secular space; the royal temple, symbolizing morality and harmony between earth and the cosmos; and the public square, which symbolizes the king as the conductor of ceremonies of state and the connection between the supernatural

---

[240] Paul Wheatley, *Nagara and Commandery: Origins of the Southeast Asian Urban Traditions* (Chicago: The University of Chicago, 1983), 4–6.

[241] Evers and Korff, *Southeast Asian Urbanism*, 82–83. Urbanism is a concept enmeshed in a web of semantic ramifications that has persisted for some five millennia as a means of denoting sets of qualities possessed by the larger, and usually more compact, clusters of settlement features that, at any particular moment, have represented centers of continuous population movements. But no city—indeed, no urban tradition—has survived for more than a fraction of the time during which the concept has been in existence. One of the first urban settlements to fall under the "urban process" of "urban generation," in the sense of it being one of the only examples of documented settlement by both archaeological and literary sources, was Fu-nan (present-day Vietnam).

[242] Ibid., 82. See also Sivarska Siulak, *Siamese Resurgence: Asian Cultural Forum Development* (Bangkok: Sukhsit Siam, 1985).

world and earth.[243] Another ritual is the swinging ceremony, which enacts processions through these spaces to connect the Brahmin temple and the main royal complex, forming the main road and axis of the city of Bangkok as well as the main path for future festivals.[244] Quite contrary to physical mapping, these ceremonial processes of engagement of the citizens and hierarchy of the city determine the sacrality of space as well as construct and define the transition of rural to urban space through ritual.[245]

Center

A blood distinction between rulers and commoners in communities determined the beginnings or original centers of urbanization for premodern and ancient Southeast Asian cities.[246] Urban areas would then flourish around the recognized ruling elite and become the point of difference between city and village. There was also a distinction between the center and the periphery for some Southeast Asian cultures, in terms of which polity would be destined as a capital and therefore would be hierarchically situated above any of its neighboring towns.[247] These ideas are prevalent in cosmographic mappings, originating in South Asia, such as the *mandala*, with Mount Meru at its center (refer Figure 5).[248] However, these mappings of concentric circular space were essentially diagrammatic and did not necessarily operate as literal translations of spatial ordering systems.

Different understandings of the notion of center occurred between Southeast Asian cultures. O'Connor suggests that for the Malays the ruler or *raja* was the most important central figure, whereas for the Javanese it was the palace (*kraton*), and for the Thais and Burmese, the city (*mųang*, *myo*) occupied the

---

243 Evers and Korff, *Southeast Asian Urbanism*, 82.
244 Ibid., 83.
245 Ibid. See also Ban Lee Goh, *Urban Planning in Malaysia: History, Assumptions and Issues* (Malaysia: Delta Editions, 2004).
246 Richard A. O'Connor, *A Theory of Indigenous Southeast Asian Urbanism*, Research Notes and Discussions Paper No. 38 (Singapore: Institute of Southeast Asian Studies, 1983), 23.
247 Ibid., 71. Also referred to as "urban-centered" social hierarchy.
248 Ibid., 53–55.See also Anthony Reid, "The Structure of Cities in Southeast Asia, 15th–17th Centuries," *Journal of Southeast Asian Studies* 11 (1980): 235–50.

center.[249] Accordingly, a settlement was not considered a significant entity as a city (*kota*, *bandar*) without the presence of a *raja*. If the residing *raja* died, tradition ensured that his palace was abandoned, and for the incoming *raja*, a new palace was built, altering the urban form of the settlement. By contrast, the traditional Javanese polity centered on the palace (*kraton*), and the palace, capital, and kingdom were represented by a single name. The whole of society was organized around this center in concentric circles. The first circle contained the capital city or *nagara*, followed by the *pasisir* or coastal provinces on the periphery. Javanese did not have any precise terms for city or town, and these concentric urban configurations were essentially the palace at the center surrounded by series of villages.[250] Thai and Burmese words for towns did not reference rulers or palaces. Typically, in the Thai case, the settlement had a capital and then four concentric cities.[251]

Geography

Geographic qualities of the Southeast Asian regions also played a large part in settlement formation. Royal cities were typically located inland because of a number of factors, but a major reason was for defensive purposes. Often, the urban form of these cities was walled, if they were stationed in highland areas. A close proximity to the coast was perceived as a threat from pirates as well as foreign invaders coming by sea. Many royal or sacred cities were also connected to belief systems that empowered sites of natural geographical features such as mountains and water. There were also exceptions, such as in the Burmese situation, where the king and dynasty held more importance than the geography. The location only became important and acted as a center of control if the king remained there; thus, shifting the capital may have occurred between dynasties.[252]

---

[249] O'Connor, *Theory of Indigenous Southeast Asian Urbanism*, 62. For the Malays, having a *raja* was the most important aspect of a state and, according to Milner, "the raja was 'the only institution', 'organising the principle of the Malay world.'"
[250] Ibid., 62–63.
[251] Ibid., 63. See also Hans-Dieter Evers, "The Culture of Malaysian Urbanisation: Malay and Chinese Conceptions of Space," *Urban Anthropology* 6, no. 3 (1977): 205–16.
[252] Evers and Korff, *Southeast Asian Urbanism*, 75.

For the case of the commercial cities, Evers refers to cities such as Malacca that existed as important nodal ports within a global trading network.[253] Malacca's geographical location on the coast, in addition to its access from inland territories for the delivery of resources, spices, and textiles from inland villages, ensured its commercial success. Perhaps, also, its multiethnic community meant that aspects of cosmology, religion, and ideology were deemed less important than economic priorities.[254] Malacca's urban form changed to represent European planning models, around the early sixteenth century following colonization by the Portuguese.

The third type of urban formation identified by Evers is the intermediate and smaller cities, which were of local importance.[255] Again, geographic location and topography were crucial because they were nodes within trade networks and stationed between royal cities and commercial cities. Some examples are Pegu and Syriam in Burma; their primary purpose was to serve the main urban center, but given that these cities contained both trade and provincial controls, they were a challenge to political power of the center.[256]

3.2 Settlement and Sacred Geography

Embraced in Southeast Asian cosmographical representations, geographical features of the natural landscape gave meaning to everyday space. From sacred principles, mountains and water acted as places of pilgrimage and refuge, and sites of festivals, which in turn guided the proximity of settlements. Physical features of the geographical landscape provided pragmatic opportunities for settlements to prosper by means of rivers or oceans for trade and transporting goods; however, for growing crops, both the hinterland and coastal regions suited these aspects of Southeast Asian life differently based on their different microclimates. Geography had significance for settlement, from the perspectives of the sacred as well as the physical. This was evident in all forms of mappings,

---

[253] Note: The Malay spelling is Melaka.
[254] Evers and Korff, *Southeast Asian Urbanism*, 30.
[255] Ibid. See also W. F. Wertheim, "Changing Southeast Asian Societies: An Overview," in *Sociology of Southeast Asia: Readings on Social Change and Development*, ed. Hans-Dieter Evers, 8–24 (Kuala Lumpur: Oxford University Press, 1980).
[256] Evers and Korff, *Southeast Asian Urbanism*, 30.

from the cosmographical to sea charts and route maps, and sacred architecture represented constructed sacred geographies.

Cosmographies

Historically, broad and diverse arrays of cosmographic conceptions coexisted with religious and cultural diversity within the Southeast Asian region. Buddhism and Islam may be considered the predominant religions, but Hindus and Christians have also had a strong presence. Space for premodern Southeast Asians thus presented a complex set of relations involving cosmographic principles, places of sacred significance, and religious concepts and symbolism. One religious practice that gave space sacred meaning was pilgrimage. Shrines and temples were plotted on maps as sacred sites to encourage visitation by pilgrims. A seventeenth-century example depicting Songkhla Lagoon situated on the Gulf of Siam indicated the eastern bank of the lagoon with sixty-three temples.[257] This Songkhla Lagoon map blended sacred geography with actual buildings on the earth's surface.

Records of traversal between earthly worlds and metaphysical realms and mappings of imaginative space are not unique to Southeast Asia. Other premodern cultures that had georeligious concepts to aid conjectures of imaginative landscapes also influenced the Southeast Asian context. Islam's concept of *fada'il*,[258] the Chinese notion of geopiety, as well as pilgrimage associated with Christian faiths, which influenced local monks and missionaries, were some techniques for constructing imaginative space.[259]

---

[257] Winichakul, *Siam Mapped*, 29.

[258] Samer Akkach, "Religious mapping and the spatiality of difference," in *Thresholds (Cambridge)*, Massachusetts Institute of Technology Department of Architecture, (2002): 73. Through the identification of certain landmarks of religious significance, the *fada'il* bestows these built forms within the socioreligious landscape with eschatological significance." In turn, a type of mapping occurs by which the *fada'il* in its narrative sets the scene and creates a particular spatial experience for faithful followers of the scriptures. Therefore, in its construction of a certain geography of divine significance, the concept of *fada'il* promotes an imagined spatiality and view of the world, authorized by God and conveyed and open only for his believers.

[259] Tuan, "Geopiety," 12–15. Georeligious belief systems were paramount in premodern Chinese and Roman cultures for determining one's place in the world. Yi fu Tuan highlights that, through the concept of "geopiety," this spiritual connection with place was inscribed in religious canons across these cultures. In turn, geopiety as a belief system being intrinsically linked to an individual's imaginative conception of space dictated every aspect of earthly life. The notion of piety covers relationships between people with the land, human beings, and their god or gods, as

In Joseph Schwartzberg's study of Southeast Asian tribal cosmographies and the Ngaju Dayaks of Kalimantan, he explains how different readings of the *Traiphum* extend ideas and imaginings of geography for premodern peoples. Ngaju Dayaks imagined spatialities of an Upperworld and Underworld, utilizing imagery such as the "village of the dead" and the "soul boat" to transport people from one form of existence to the other. Priests used imagery as cues in their chanting, to aid transition from one existence to another. Females descended to the Underworld and males ascended to the Upperworld. Successful transitioning was achieved over a designated period of several nights, with priests chanting to end the respective journeys.[260]

Hans Schärer, a Swiss Protestant missionary who studied the Ngaju Dayaks in the 1930s, described the imaginative geography of the Upperworld. Among other features, this geography, exhibited the primeval mountain—domain of Mahatala, supreme deity (a hornbill)—giant Trees of Life laden with bountiful fruits, and everything found in the Middle world (human world), but of greater and more beautiful proportion (refer Figure 6). By contrast, the Underworld or "primeval waters" was presided over by the Goddess Jata (a water snake).[261] Access was achieved to the Underworld at the joining of rivers and their deepest sections of water. Imagery of this realm included fish caught in fish traps—the food source of the Upperworld—as well as Jata's crocodile attendants (refer Figure 7). As Ngaju Dayaks traverse between worlds of existence from physical space to their sacred village to worlds of other nationalities, their ritual behaviors also change.[262] Perhaps this demonstrates the close connection between imagined, sacred geographical space and physical repercussions of cosmological experiences for the Ngaju Dayaks. As a highly localized version of passage between life and death, the Ngaju Dayaks' practices reflect the variety of interpretations originating from Indian cosmology portrayed in mappings.[263]

---

well as human beings with nature. *Hsiao*, the Chinese term for piety, translates as "filial piety" or "filiality"; hence, the duty of care for one's parents and ancestors is a dominant aspect of pious Chinese.

260 Schwartzberg, "Cosmography in Southeast Asia," 703.

261 Ibid., 704.

262 Hans Schärer, *Ngaju Religion: The Conception of God among a South Borneo People*, trans. Rodney Needham. The Hague: Martinus Nijhoff (1963):64–65 (note 5), cited in Schwartzberg, "Cosmography in Southeast Asia," 710.

263 Schwartzberg, "Conclusion to Southeast Asian Cartography," 840.

Views of the world for premodern Southeast Asians, and hence mappings, incorporated a representational merging of physical, metaphysical, and spiritual realms. An example in Thai Buddhism is the painted bark-paper manuscript "The Four Continents Around Sumeru." Its illustrations conveyed a sense of the close relationship between earthly and non-earthly existence in Thai culture. [264] The *Traiphum*, or "Three Worlds' Cosmography," dating from the fourteenth century—perhaps the most referenced Buddhist cosmological series—is significant in Thai culture because it depicts premodern spatial conceptions as well as resonates and influences ideas about time and space in modern times.[265] Thongchai Winichakul, in his book *Siam Mapped*, notes the paradoxical nature of the text:

> Space in the *Traiphum* was the qualitative manifestation of imagined existence. Yet all of the surviving *Traiphum* texts give concrete descriptions of various worlds, particularly the human one, as well as accounts of the movements of the sun and moon and the seasonal changes.[266]

It is debatable to what extent the *Traiphum* was intended to convey a sense of "real-worldliness." In the case of the *Traiphum*, "if one concedes that a map does not have to be a representation of the earth's surface, but can depict other relations of space, it is easier to understand this set of pictures [dated from 1776, the manuscript has no title, but the one given to it later, *Samutphap traiphum chabap krung thonburi* (Pictorial Manuscript of the Traiphum: Thonburi Verison)]."[267]

Versions of the Three Worlds Cosmography, according to Day and Reynolds, in its initial compilation, resulted from more than thirty Buddhist texts

---

[264] "Its composition included Mt Sumeru where earth was perceived as a horizontal disk with Mt Sumeru at the centre … the seven surrounding mountain ranges and seas, the four principal continents inhabited by humans, each with its distinctive shape, and multiple tributary continents nearby." Schwartzberg, "Cosmography in Southeast Asia," 725.
[265] Tony Day and Craig J. Reynolds, "Cosmologies, Truth Regimes, and the State in Southeast Asia," *Modern Asian Studies* 34, no. 1 (2000): 8.
[266] Winichakul, *Siam Mapped*, 20–21.
[267] Ibid., 26.

and accredited as the oldest work of Thai literature.[268] The text was again reworked against "Pali sources from 1783 to 1802 by the order of Rama I."[269] By contrast, most of the surviving manuscripts that date from the sixteenth century contain very little text. Over subsequent centuries, the *Traiphum* addressed ideas about the geography of the time and evolved under changing scientific as well as religious ideas. Graphic interpretations of the *Traiphum* text were painted over the walls of Thai monasteries by the nineteenth century, with the intention of spreading both the "faith and knowledge" of the studies of Theravada Buddhism.

Variations of conceptions of *Traiphum* space were evident in its numerous artistic impressions. According to Winichakul, differences of interpretation were largely reflective of the school of painting rather than altered concepts of space or extent of geographical knowledge.[270] There are instances in the *Traiphum* narrative in which earth geography and *Traiphum* space relate, although they are different. For example, the inclusion of the coastal map in the *Traiphum* based on Chinese coastal charts, indicating indigenous knowledge about the earth's surface, was not always removed from scientific information.[271] The coastal map related to an actual geography, rather than cosmological space, in its depiction of places along a coastal itinerary. Imagery of the *Traiphum* also was intended to educate its viewers about the social order of Thai society as well as portray the cosmos. Rama I played an instrumental part in reorganizing aspects of Thai social order evident in earlier versions of the manuscript, by relocating humankind to the middle order. Earlier manuscripts conveyed different presupposed levels of existence, from lowest to highest. As the human representative of divinity, Rama I justified his reordering of the text, fostering the Buddhist moral order and exerting control over imaginative space of the *Traiphum*.[272]

Highlights of these preserved sections of the *Traiphum Thonburi* manuscript include illustrations of the three worlds, representing an imaginative geographical interpretation of Southeast Asia and the Indian Ocean.[273] The physical form of the manuscript is a thick indigenous bark-like paper, which is

---

[268] Schwartzberg, "Cosmography in Southeast Asia," 720.
[269] Day and Reynolds, "Cosmologies, Truth Regimes, and the State," 8.
[270] Winichakul, *Siam Mapped*, 24–25.
[271] Ibid., 30. See also J. V. Mills, "Chinese Coastal Maps," *Imago Mundi* 11 (1953): 151–68.
[272] Schwartzberg, "Cosmography in Southeast Asia," 721.
[273] Ibid.

stuck together in an "accordion-style" format.[274] The Thonburi manuscript with all 272 leaves stretches to 50.9 meters.[275] If read from left to right, it indicates the sacred Buddhist Indian temples, the Bo tree, believed to be Buddha's birthplace, then his mother's town, followed by the Bo tree under which he achieved enlightenment. This part of the map indicates many sacred sites within close proximity of one another. As the viewers' eyes traverse the map to the right, there is an element of transition in descending from the metaphysical to earthly realm of Siam. At the most central point of the map, northern Thai towns such as Chang Mai and Sukhothai are depicted. Again traveling to the right, more towns are indicated, until Bangkok and the southern peninsula is reached; the graphics of the map then change to record timelines of days between towns. To the west, serpents surface in the Narai River near Rangoon and Burma. "They ask Buddha to leave his footprint upon the land, which would be a holy and fortuitous occurrence."[276] At the bottom border of the map, where the monkey (aka King Rama's knight) is indicated overlooking islands, there is land linking the large island that is Ceylon, which recalls the legend of the Ramayana. The Ramayana is an Indian epic, in which King Rama goes to rescue his captured wife held in Ceylon and runs along this land bridge to reach her. Buddha also used this bridge to achieve enlightenment, leaving his footprint on Adam's peak, also indicated here. The island conveying two naked people is said to be the Nicobar Islands, although not located according to true geography, that is, north of Sumatra.[277]

Winichakul qualifies that this is not an inaccuracy or misjudgment of the mapmaker, but rather the *Traiphum* records places in accordance with the "description about Buddhism in Suwannaphum," not relative to the earth's surface.[278] Thus, cosmological messages of the *Traiphum* triumph over "actual" geographic information within the manuscript.

Other ways in which indigenous thoughts on geography incorporated notions of the transcendental, the physical, and the empirical aspects of life were

---

[274] Ibid. See also Frank Reynolds and Mani B. Reynolds, *Three Worlds According to King Ruang: A Thai Buddhist Cosmology* (Berkeley: University of California, 1982).
[275] Schwartzberg, "Cosmography in Southeast Asia," 721. See also Jean Boisselier, *Thai Painting* (New York: Kodansha America, 1976).
[276] Suárez, *Early Mapping*, 42.
[277] Ibid.
[278] Winichakul, *Siam Mapped*, 27.

through conceptions of temporality.[279] Maps were often considered mediums of impermanency to fulfill short-term objectives, be they for religious ceremonies or invasion strategies. For example, Buddhist *mandalas* demonstrated the belief in the impermanence of life because they were painstakingly created in elaborate detail and then swiftly erased after their role was fulfilled in one ceremony.

Sacred Architecture

Sacred building complexes not only designated types of settlements but also informed their citizens of sacred practices and proposed expectations for citizens within the society. These designations served as sociocultural and religious mappings, as well as teachings, of the role one should play in society, while also laying out the citizen's destiny as long as they subscribed to cultural rules. Former temple complexes, of the early ninth century, have been found in Javanese archaeological records associated with the *Śaivite* dynasty. These temples were evocatively set amidst a landscape of sulphur lakes and volcanic craters. Stone staircases that enabled worshippers to access the temple as well as foundations of nearby temples have been found intact.[280]

"Indianized" areas of early Southeast Asia consisted of residents directly from the subcontinent and therefore "served to confirm and consolidate the emergent notion of western Southeast Asia as an integral part … of the South Asian, principally Hindu, cultural milieu."[281] In terms of a settlement pattern, the result was a well-defined urban morphology as well as a relatively well-functioning, internal organization and quality of life, commensurate with a "new level of socio-cultural integration."[282] The Khmer temple Angkor Wat represents a sacred, constructed three-dimensional mapping and representation of space and time, demonstrating ancient and premodern principles of spatial planning.[283] If we consider Angkor Wat Temple a representational medium, we understand that a

---

[279] Suárez, *Early Mapping*, 27.
[280] Wheatley, *Nagara and Commandery*, 238.
[281] Ibid., 420.
[282] Ibid.
[283] To return to the notion of representation with regard to settlement, another definition for the former coined by the *Compact Oxford English Dictionary* is a "material image or figure; a reproduction in some material or tangible form."

map, by definition, may be broadly conceived, not as a representation of the earth's surface, but as a depiction of spatial relations.[284]

Angkor Wat temple, as a three-dimensional exhibition of complex Indic cosmological design elements displaying detailed documentation of cosmological painted imagery within its walls, impressively constructs another type of spatiality. As they traverse the temple site, visitors are rendered participants in the Golden Age of Indic cosmology. Traveling certain paths within the complex, as well as immersion in artistic portrayals of cosmological events, encourages engagement at both spatial and spiritual levels. The divinity of the king, who presided over the construction of the temple, is prioritized in the design using seasonal daylight to illuminate statues, suggesting a royal presence within the building. Eleanor Mannikka, in her book *Angkor Wat: Time, Space and Kingship*, describes an imaginative geography where the notion of time is played out in the visitor's spatial experience of this twelfth-century Khmer temple. Distances traversed at the entrance and then upon leaving the central temple equate to 1,728 cubit lengths, which also relates in time measurements to the duration of the Golden Age.[285] Therefore, a complex set of relations exists within the temple grounds, between spatiality and time as well as architectural form and metaphor. Sacred Mount Meru and ascension to a higher cosmological plane are also referenced in the physicality of the tallest central tower.

This temple represents a glorification of the reign of King Suryavarman II, who commissioned it, and demonstrates the journey into another kind of existence, a metaphorical construction of an imaginative time and space. Temple design uses daylight at particular times of the day for maximum effect to achieve different moods within constructed allegorical space. As an example, the passage to immortality regarded as the Churning requires all subjects of King Suryavarman II to live in harmony. The rising sun emphasizes the Churning at the beginning of the day, and as seasons change and light diminishes, alternate deities are illuminated on the temple walls. The clock of nature dictated time in the design. Winter casts a shadow over gods and highlights the king, and the reverse occurs in summer. However, it is a combination of all knowledge systems

---

[284] Winichakul, *Siam Mapped*, 26.
[285] Day and Reynolds, "Cosmologies, Truth Regimes, and the State," 5.

integrated into the temple design, from its orientation to its adherence to astronomical patterns that delivers its powerful impression upon scholars and believers. Physical form also symbolizes cosmological ascendancy in the temple. The central tower is the representation of both the central peak of Mount Meru and the home of God, Visnu. To reach Visnu's house, a climb of equivalent distance to thirty-three units also reflects the universe as a whole.[286]

> In complementary terms, the total vertical length of all four corner towers … symbolizes a complete *mahayuga* period of 4,320,000 years. As a consequence, the four towers arranged around the central tower can symbolize both the real cycle of the stars and planets around the north pole as well as the four *yuga* cycles … As the north celestial pole, Mt. Meru remains immobile while time forever flows around it in an eternal movement.[287]

However, for premodern Southeast Asians, local territory was familiar and to a certain degree assumed knowledge of, as well as its relationship to, other kingdoms. The greater geographical and astronomical picture was an issue for the realm of cosmology. Therefore, different kinds of space coexisted within premodern geography, and similarly, an infinite number of messages may be derived from such representations of space. The varieties of messages are linked to the knowledge base of the viewer; therefore, the process of viewing enables a shift from the imaginary to the real. Indigenous views of space linked to the sacred and profane permitted meaning in the act of conceiving of space. Spatial concepts were appreciated and understood based on sets of knowledge principles, whether they were religious or cosmological. Both the *Traiphum* and Angkor Wat provided two different realms in which to explore these georeligious spatial concepts, and gave the participator certain theoretical tools and sets of relations with which to negotiate sacred, constructed, and physical geographies.

---

[286] Ibid., 6.

[287] Eleanor Moron, "Configurations of Time and Space at Angkor Wat," *Studies in Indo-Asian Art and Culture* 5 (1977): 249.

3.3 Settlement and Natural Geography

Natural geography and settlement in premodern Southeast Asia were intimately connected for different peoples. The relationship was often cosmological, but the origins of this thinking also reflected the pragmatic need for human survival. Therefore, settlement positioning, although cosmologically conceived, was ultimately driven by practical needs. Coastal and inland regions were the two key areas of the most significant forms of settlement, and intermittent smaller villages stationed between them.

The physical nature of these two types of cities has been touched upon, but noted here is the significance in settlement formation of the natural components of water and mountains. First, if we further examine the royal city of Angkor Wat, we find the connections between water and mountains are cosmological but intertwined as a required part of the social and religious structure of the city for the ancient Khmer population. Inland royal cities were able to recreate the significance of these natural elements of life within their urban forms, in symbolism and for pragmatic purposes. Second, coastal cities, such as Bantam in the Sunda Strait, were seen by the Dutch traders in the sixteenth century as useful ports for access to the markets of Java, and were a key source of spices.[288] The ocean location of Bantam in the two images provided prioritizes the sea view because this was the focus of colonizers as well as the local populations.

Water

Angkor Wat demonstrates the design of sophisticated irrigation networks within its grounds.[289] Ancient India featured water in its cosmologies; similarly, at Angkor, a series of moats around the temple serve to place the temple in a symbolic "ocean of cosmic creation."[290] There are also temple tanks and lakes, which also relate to this notion of the cosmological importance of water. The temple of Neak Pean reflects the heightened cosmic relationship between water and building because it is set on circular terraces, in a square water tank with

---

[288] Suárez, *Early Mapping*, 191.
[289] Michael W. Meister, "Mountains and Cities in Cambodia: Temple Architecture and Divine Vision," *International Journal of Hindu Studies* 4, no. 3 (2000): 263. See also A. K. Coomaraswamy, *Yaksas: Essays in the Water Cosmology*, ed. P. Schroeder (Delhi: Oxford University Press, 1993).
[290] Meister, "Mountains and Cities in Cambodia," 263.

water flowing beyond into a series of pools and irrigation channels.[291] Groslier observes:

> This tower-sanctuary, representing paradise floating on the primaeval ocean, rises on circular foundations in the middle of a square basin. Water from the main basin flows down to four smaller tanks round from fountains in little shrines. It is the symbol of a miraculous lake in the Himalayas where all illnesses were cured ... The waters of Neak Pean, judiciously redistributed by canals, flowed towards the four points of the compass and there transformed the rivers and liquid arteries of Angkor into as many magic streams.[292]

These canals' earthly purposes included irrigating the Khmer population's fields and sources of food, while symbolizing the recreation of a celestial city on the ground. The physical nature of Bantam's settlement has not been well documented, but engravings suggest an established settlement. The Chinese source, *Chu-fan-chi*, written around the thirteenth century, highlights that Srivijaya still ruled Sumatra, the Malay Peninsula, and western Java in the Sunda Strait. The source identifies that the people worked in agriculture, and their houses were built on wooden poles (*rumah panggung*).[293] If this was the case, the account suggests a different building typology between towns on the coast and those of the hinterland in this region. The significance of water in the connection between hinterland and the sea is also evident in the illustration below.

Mountains

Mountain geography was crucial to conceptions of Angkor Wat, first, because it was situated in the hinterland away from the coast, but it also incorporated the symbolism of sacred mountain geography into its constructed

---

291 Ibid.

292 Bernard Philippe Groslier, *Indochina: Art in the Melting-Pot of Races* (London: Methuen, 1962), 181–82.

293 R. Soekmono, *Introduction to the History of Culture Indonesia*, 2nd ed. (1973, New York: Canisius, 1988), 60.

space. Mount Meru, the central mountain of the sacred schematic of Indian cosmology, was a part of the spiritual cosmological mapping of Angkor Wat.[294] The central temple complex of Angkor was where the fertility of the earth and spiritual powers joined. Mountain geography was important in every aspect of Southeast Asian life as a primary form of the physical and spiritual associations with landscape. From the sovereignty of the king, mountains were linked, in physical and symbolic terms, to every aspect of life.[295] "Mountains were the source of spiritual life; they were also the origin of the other 'element' of the world, rivers, and thus were the source of physical life as well."[296]

The connection between mountainous geography and water sources is also evident in maps of the Sunda Strait and the township of Bantam. Although coastal areas are typically near to sea level and characterized by plains, depictions of the landscape show hills bordering townships. Even though a scale is indicated, it is difficult to obtain a sense of the topographical relationships. However, images of Bantam do suggest a clear connection and proximity of a river system with the mountain range beyond, thus reinforcing the importance of these aspects of natural geography to settlement location and formation.

3.4 Settlement and Representation

Illustrations and maps relay specific qualities of the built environment, and these may be urban or non-urban, yet their contribution is instrumental to the holistic impression of the place. As we have seen, premodern maps perform a visual reporting role on integral aspects of everyday life, from the religious and mythical, to determining sanctity, routes, and patterns, all of which affect settlement. Our understandings of premodern maps have shifted to embrace three-dimensional mappings and sacred architecture as a means of conveying spatial concepts, time, and space relevant to premodern Southeast Asian societies. Yet many of the premodern geographies presented in this study display a malleable quality whereby each new narrative enables a reinvention of the previous geographical conception. Premodern maps tell us about critical aspects of human environments, about belief systems and cultural values.

---

[294] Suárez, *Early Mapping*, 24.
[295] Ibid., 34.
[296] Ibid.

Regalia

In premodern Southeast Asia, Malays and Javanese peoples viewed royal regalia as denoting an important person. For example, if the regalia was removed, so was the person's and the city's significance. Regalia may be a map of sacred significance, as outlined below. In contrast, Thais and Burmese focused their concerns on Buddhist relics and images, which were city centered. Royal regalia, such as weapons, musical instruments, and royal insignia, would protect the polity and bring it prosperity. Buddhist relics and images existed as passive entities in the overseeing of operations of the Thai and Burmese cities.[297] Often, cities were imagined or associated with their most outstanding relic or religious monument. Thus, irrespective of who ruled the city, the image of the relic related to both the current ruler and the city.[298] In Thailand and Burma, a *coup d'état* would involve seizing the king's palace, not the relic or image associated with the city. In the Indonesian case, sacred regalia, if seized, would assert the thief's status and confirm the thief's self-pronouncement of king. By the nineteenth century, when there were a number of landless or displaced *rajas*, and rulers lost their kingdoms, the royal peoples themselves were still important despite their uprootedness and loss of a significant physical place within society with their regalia intact. By comparison, if the Burmese kings lost their palace, they would lose their status, as well as any previous recognized power.[299]

"Central Portion of the Sacred Map of Timbanganten" is an example of a "sacred relic" or *pusaka* and is housed in Ciela, a village of the Garut District from the Sundanese region of Java (refer Figure 8).[300] The image indicated here is only one portion of the map, which is a large work, detailed in its construction on a cloth background. Ciela is located sixty-five kilometers southeast of Bandung, at the base of a volcano, Cikuray. Timbanganten refers to the minor seat of the

[297] O'Connor, *Theory of Indigenous Southeast Asian Urbanism*, 34. See also Robert Heine-Geldern, *Conceptions of State and Kingship in Southeast Asia* (Ithaca: Cornell University Southeast Asia Program, 1956).
[298] O'Connor, Theory of Indigenous Southeast Asian Urbanism, 64; F. H. Naerssen, "Tribute to the God and Tribute to the King," in Southeast Asian History and Historiography: Essays Presented to D. G. E. Hall, eds. C. D. Cowan and O. W. Wolters (Ithaca: Cornell University Press, 1976).
[299] O'Connor, Theory of Indigenous Southeast Asian Urbanism, 65–66.
[300] Schwartzberg, "Southeast Asian Geographical Maps," 767.

chiefdom located in the village of Ciela, which existed prior to the breakup of the Sundanese Hindu kingdom of Pajajaran in the 1570s.[301] The Sultan of Bantam, who had embraced earlier in the 1500s, instigated the break up, and force was used to convert all in the chiefdom to Islam. The beginning of the Javanese people's conversion to Islam did not eventuate until 1525. The mapmaker recorded his connection with a prince "*Sunan Lawas Jaya*," who is unknown in history, and was a Muslim who commissioned the map to determine the extents of his kingdom.[302]

Text on the map indicates that the kingdom comprised seventy-eight *kampongs* (villages), which were distributed first into three sections and then subdivided further. The orientation is to the south, although not consistently adhering to that direction; it also attempts to cover approximately 40,000 square kilometers, but actually indicates the chiefdom of Timbanganten at an exaggerated scale. The notable qualities are the cartographic aspects relating to the actual geography of that part of Java, such as its interpretation of drainage features and topographic patterning of the landscape.[303] Given the map's sacrality, it is kept under the surveillance of a *kuncen.* A widely respected person in the village holds this position.[304] Holle explains in his account of experiencing the map:

> That this map is greatly revered can be seen on Friday eve. Then, the *kuncen* unwraps it so the population can view it. Then, a brazier is lighted with incense, and the *kuncen* mutters an Arabic prayer, which only she is allowed to pray, and which she repeats continuously. The population lingers around the house, taking in eagerly the sounds of prayer, although it is inaudible to them, and watch with equal respect the sacred cloth which is equally obscure to them. The thought that it once belonged to their former rulers,

---

[301] Ibid., 768. See also O'Connor, *Theory of Indigenous Southeast Asian Urbanism.*
[302] Schwartzberg, "Southeast Asian Geographical Maps," 768.
[303] Ibid., 769.
[304] Ibid. See also Charles A. Fisher, *South-east Asia: A Social, Economic and Political Geography* (London: Methuen, 1964).

> and that it was even made by a powerful *dalem* [ruler], leaves them in their worshipful mood.[305]

Paradoxically, regalia from Malay and Javanese cultures heightens the importance of the role of the map if it happens to have sacred significance, but diminishes the priority of settlement tied to a particular place.

Patterns

Intrinsic to spatial organization, patterns shape understanding of concepts from cosmological concerns to human behaviors. For premodern Southeast Asians, patterns demonstrated by cosmographies defined time, space, and seasonal variations in agricultural practices in addition to principles of sacred architecture. Patterns are synonymous with ordering principles and practices of Southeast Asia such as feng shui, mandala diagramming, *fenye* agricultural systems, and tectonic symbols of the sacred, which are necessary to visions and creations of settlement.

The patterning employed in a batik map is very dense; the main motif, which is a series of squares with three-pointed roofs, is believed to be the Puri gates or traditional entrances to Javanese Hindu temples.[306] The map does not display any text, and the date, creator, and geographical context depicted are unclear.[307] It is a colorful composition in pale blue and orange, with gold foil inlay against a background of dark indigo. The rectangles within each of the square motifs are thought to be representative of villages, with wavy patterns around them, which may symbolize either world oceans or the lakes around Mount Meru. The vast majority of the intense line work may represent rice fields and terraced agricultural patterns practiced in Indonesia, which also implies ascending and descending sets of contours to the landscape.[308] Possibly, this map was produced to represent a traditional Hindu Javanese pattern known as the *Moncå-pat*, discussed in 1918 by Dutchman van Ossenbruggen:

---

[305] Schwartzberg, "Southeast Asian Geographical Maps," 770.
[306] Ibid., 772.
[307] Ibid., 771.
[308] Schwartzberg, "Southeast Asian Geographical Maps," 771. See also Daniel George Edward Hall, *A History of South-east Asia*, 3rd ed. (New York: St Martin's Press, 1968).

> *Moncå-pat* in middle Java points to a unit of a desa [locality] with its four neighboring *desas*, which one should think of as being arranged according to the four cardinal points. This unity of *desas* extends further, including even more distant areas, and is connected with the old Javanese system of unity of which, even today, traces can be found in the law of the princely states.[309]

There is no direct evidence to suggest that these spatial relationships were reconstructed in later evolutions of settlement in Indonesia or Java, but the principles perhaps had greater application in an earlier era. This system of ordering village space was to give villagers a sense of ownership and thus responsibility to prevent crime and disorder. Villages were organized outwardly in the cardinal directions from the main capital. Over time, such systems of control broke down with land expansion. Although highly conjectural, this possibility of an urban patterning system is among the only examples of structural urban space conveyed in Southeast Asian maps.

Journey

Graphic representations of journeys, through cartographic signs, revealed many unknown aspects of space and culture to premodern peoples, including distances between villages, urban expanse, and numerous other relationships of comparison with peoples within close proximity. The predominant group of non-cosmographic maps fall into the following categories: route maps, large-scale maps of reasonably small areas, large-scale maps of cities, and at a larger scale still, architectural plans, typically of temple architecture and other prominent city buildings. Each of the following maps demonstrates attributes that are important for settlement, from connection between geographical elements to relationships between places and cultural groups.

---

[309] F. D. E. van Ossenbruggen, "De oosprong van hat Javaansche begrip Montjå-pat, in verband met primitieve classificaties" [The origins of the Javanese concept of *Moncå-pat* in connection with primitive classifications], *Verslagen en Mededeelingen der Koninklijke Akademie van Wetenschappen, Afdeeling Letterkunde*, 5th ser., pt. 3 (1918): 6–44, esp. 6., quoted in Schwartzberg, "Southeast Asian Geographical Maps," 772.

"Map of the Vale of Manipur showing the route of King Alaungpaya's invasion of 1758–59" possibly indicates settlement patterns of the different ethnic minorities of the area through different coloring codes. For example, Burmese settlements are yellow, and Manipur settlements are represented by the use of lavender or tan colors. Therefore, this suggests that lavender may represent Hinduized Manipurs as opposed to the yet-to-be-converted Naga tribes. There is a dominance of pagodas (more than a dozen) indicated on the Burmese side, whereas there is only a single Hindu temple indicated in Manipur to represent the capital Imphal.[310] This Shan Burmese map offers a detailed interpretation of Manipur, as well as the northeastern neighboring region of India and some parts of northern Burma. It is most probably Indian muslin and comprises two pieces that were sewn together. Most likely, this map was drawn from hearsay.[311] For the Burmese, geographic information was accrued for revenue surveys and organized census activities.[312]

Route maps provide a context for each settlement in that they define their isolation or close proximity to neighboring cultural groups. In the "Map of the country of North Ava," an important aspect of the map is that there is a record of distances between places. These distances are noted either as days of travel or in a measure of Burmese leagues (*datn*), which are equitable to 2.2 British miles, enabling an observer of these maps to establish a degree of scale. An intervention by Francis Hamilton may be indicated by the dashed ruled lines, which refer to the routes between places.[313]

The main features of an 1861 map "Small excerpt from a large Burmese Map of much of the Eastern Shan States" are the details and efforts expended in representing the mountain ranges and vegetation for environmental management; notes in Burmese on the map state "woodcutting forest" as well as "woodmaking forest."[314] The notes accompanying the map go so far as to state that the woodcutters should come from a specific town. Other cartographic developments

---

310 Schwartzberg, "Southeast Asian Geographical Maps," 754.
311 Ibid., 753.
312 Hall, *History of South-east Asia*, 585.
313 Schwartzberg, "Southeast Asian Geographical Maps," 748.
314 Ibid., 756.

include significant and insignificant watercourses noted as well as two different units of measure for distances between towns.[315]

3.5 Concluding Remarks

Premodern mappings for Southeast Asians illustrate specific sociocultural ways of seeing the world. However, maps only occupy part of the narratives, of which there are many, in the conception of settlement. Unlike modern spatial relations, Southeast Asian understandings of human environments extend beyond the limitations of the map. Practices such as sacred regalia determining persons of power rather than places of power highlight that physical settlement means different things to different peoples. The idea of journey mappings constructing spatialities relative to one another and thus the concept of settlement has meaning relative to one's neighbors. Patterns in sociocultural practice and behavior, in both earthly and spiritual realms, deliver certain sets of ordering principles to human environments. All of these qualities are dynamic in their relation to settlement geography. The emergence of modern science prompted desires for making sense of geography and a more literal translation of spatial concepts.

---

[315] Ibid. See also Rachmat Kusmiadi, "A Brief History of Cartography in Indonesia" (paper presented at the Seventh International Conference on the History of Cartography, Washington, DC, August 7–11, 1977); P. D. A. Harvey, *The History of Topographical Maps: Symbols, Pictures and Surveys* (London: Thames and Hudson, 1980).

PART II: MAPPING AND MODERNITY

CHAPTER FOUR
Mapping and Enlightenment

Late medieval and early modern mapping reflected a transition in thinking from a predominantly religious outlook to a combination of scientific and religious concerns. Developments in surveying and mapping techniques throughout the early modern period aided this transition in the Western world. By the eighteenth century, further refinement of cartographical processes and technologies for the production of surveys, route maps, and settlement planning signified a heightened requirement for geographical realism. Forms and techniques of representation reflected the aspirations whereby the city view was superseded by the ichnographic plan, making way for our modern way of depicting settlement. Despite the changes in representing and viewing the world, a recurring debate continued about religious views and how to reconceive paradisiacal space, given the geographical discoveries of the Enlightenment Colonial ventures meant new territories could be exposed to new ways of mapping and thinking about the world, put forward by the projective and rational view of the Enlightenment.

4.1 In Transition: Old World, New Vision

Transitioning to Enlightenment conceptions of space and representations was a gradual experience. Many drawing and mapping techniques and shifts in conceptualization of space and geography were evident in the medieval period. A few key aspects of medieval origins that reflect the changing nature of mapping processes leading into the Enlightenment are concepts of time and space, religious narrative versus scientific thinking, and the emergence of the survey.

Time and Space

Medieval maps often conveyed historical events that had occurred chronologically in time, but vignettes of events, such as great battles, were often arranged within the same frame as if they had occurred sequentially. Therefore, medieval illustrations indicated static landscapes in which time stood still, and different events were shown as occurring within the same scene, which created a conflict between time and space.

David Woodward, in his essay "Reality, Symbolism, Time, and Space in Medieval World Maps," suggests that these were some of the ways in which medieval mapping media were compromised in terms of their perceived inaccuracies. Mapmakers of the medieval European world intended their works to demonstrate "blended concepts of both time and space as a context for understanding the Christian life."[316] Therefore, medieval maps may be considered objects of merit in their record of geographical data as well as vestibules of multiple layers of historical ideas and events. The large corpus of medieval world maps or *mappaemundi*, which includes an extensively diverse extent of material, have both technical and ideological constraints in their resultant forms. Any major divergence of design in *mappaemundi* production was restricted by the projection of Christian historical traditions within a geographical framework.[317]

According to Jeremy Black, in his book *Maps and History*, this sense of placing events in a geographical context and time frame was not appreciated in Europe until around the fifteenth century. Black recounts another example of this shift in thinking: although maps were not usually included in storytelling, Benedictine monk Thomas Elmham used a plan of the Isle of Thanet to illuminate a medieval legend about receipt of lands by St Augustine's monastery.[318] This

---

[316] David Woodward, "Reality, Symbolism, Time, and Space in Medieval World Maps," *Annals of the Association of American Geographers* 75, No. 4 (1985): 511. See also A. Holt-Jensen, *Geography: Its History and Concepts* (London: Harper and Row, 1980).
[317] Woodward, "Reality, Symbolism, Time, and Space," 514; Hindman, *Text and Image in the Fifteenth Century*.
[318] Black, *Maps and History*, 6. See also Tyacke, *English Map-Making*.

development meant mapping was used to enable the past to be differentiated from the present.

The Strasbourg edition of Ptolemy's *Geography*, published in the early sixteenth century, ensured printed distinction between ancient and modern maps, and verified the need for addressing historical time. This reconsidering of Ptolemy's ancient texts also contributed to the already changing concepts of space around the sixteenth century. Ptolemy's theories proposed methods for representing a three-dimensional and spherical earth. Black believes this changing notion of history in the late sixteenth century led to the beginnings of what he describes as "the emergence of a quasi-modern secular historical awareness, deriving from both the Renaissance's sense of typology and progression of historical eras (Classical–medieval–modern)."[319] These events in Europe were also linked to religious changes in thinking. The divide in the Christian church led to distinguishing religious thought, before and after the Reformation. Black suggests that the main shift in translation of ideas, and thus representation of scriptures and the classics, was an emphasis placed on the literal rather than the allegorical.[320]

Historical maps were increasingly published as parts of texts, even though they did not reflect all of the developments in cartography driven by the age of European exploration. Writers of history, such as John Norden (1548–c. 1625) and William Camden (1551–1623), used maps to convey contextualization of place, indicating place-names and nominal distances, which resulted in the first forms of historical atlases. The importance of linking maps to historical narrative indicates a desire by both writers and readers of history to relate time and place. [321] Black observes, in the European case:

> The relationship between time and space is complex, but there are signs of important shifts in interest and perception in the seventeenth century … Book readers sought a history informed by

[319] Black, *Maps and History*, 6.
[320] Ibid., 7. See also Burke, *Renaissance Sense of the Past*; John Rennie Short, *The World Through Maps: A History of Cartography* (Toronto: Firefly Books, 2003).
[321] Black, Maps and History, 7. See also S. Mendyk, Speculum Britanniae: Regional Study, Antiquarianism and Science in Britain to 1700 (Toronto: University of Toronto, 1989).

> a precise cartography and a cartography that was in turn historically accurate.[322]

Previously, in the absence of any more precise renditions of urban context, generic pictures of places, such as popular views of Paris or London, had been associated with historical narratives. However, for the reader of history, these changing notions of time and space meant such generic depictions were no longer appropriate.

By the seventeenth century in Europe, relating textual accounts to place took on a new relevance, but there was also a shift in conceptions about the notion of time. Two main understandings of time resonated for Renaissance theorists. The first was Copernicus's ideas on terrestrial motion and the earth's rotation around the sun, which was the scientific theory to explain night turning to day, days of the year passing, and so on. The second was the idea of psychological experience and time not necessarily being connected to spatial movements of earth, as cited in John Locke's "Essay Concerning Human Understanding" published in London in 1690. Locke's work hypothesized that time was related to "a psychological experience of duration."[323] In terms of historical depictions of place, maps constructed a space of the past, in which time could be read differently and separately to the present. In addition, ways of seeing and reading these graphic representations were altered by developments in optics. New understandings in ways of seeing the world, in an artistic and cartographic sense, such as seventeenth-century Dutch scientific innovations, were expressed through media of paintings and maps.

Due to developments in ways of seeing, there was in turn an emphasis for Dutch cartography to convey a sense of geographic reality. Therefore, imagery associated with historical narratives shifted, from being mere decoration and the "conjuncture of time and space" to separating these two components. This affected a sense of the past and motivated new historical cartographies.[324]

---

322 Black, *Maps and History*, 7–8. See also P. D. A. Harvey, *Maps in Tudor England* (London: University of Chicago Press, 1994).

323 Black, *Maps and History*, 8. See also Alpers, *Art of Describing*.

324 Black, *Maps and History*, 9. See also Catherine Delano-Smith and Elizabeth Morley Ingram *Maps in Bibles 1500–1600: An Illustrated Catalogue* (Geneva: Librairie Droz, 1991).

Religious Narrative

The spiritual narrative overrode issues of geography or place, such as the need for an "accurate" or precise record of geographic elements. The prime responsibility of *mappaemundi* artists was to emphasize religious teachings over and above other knowledge systems. Quite by contrast, to representations of geography, spiritual histories were carefully organized and displayed chronologically from Creation to the Last Judgment.[325] Woodward points out "the logical inseparability of history and geography is thus vividly illustrated by *mappaemundi*."[326] While the notion of sight had cartographical ramifications, it was also emphasized from a religious perspective. Sight enables God to reveal his ideas about creation and humankind; therefore, reaching his believers was achieved through graphic renditions of Bible scenes.

The age of exploration and testimonies of firsthand encounters then also began to bring home archaeological evidence of the existence of human life, which contradicted the religious view. Scientists who chose to follow this route of thinking were accused of atheism, supporting witchcraft, skepticism, and adhering to pre-Adamite heresy.[327] For example, Walter Raleigh and Thomas Hariot were ostracized and arrested when they investigated the origins of Indian peoples in Virginia.[328] David Livingstone highlights how contentious relations between religious beliefs and geographical exploration resulted in a strand of geography during the seventeenth and eighteenth and into the nineteenth centuries, known as physico-theology.[329] A number of writers embraced this subject, and their views were amalgamated and emancipated in the publication *Trace on the Rhodian* by Clarence Glacken.

---

325 Woodward, "Reality, Symbolism, Time, and Space," 515. See also W. R. Tobler, "Medieval Distortions: The Projections of Ancient Maps," *Annals of the Association of American Geographers* 56, No. 2 (1966): 351–60.
326 Woodward, "Reality, Symbolism, Time, and Space," 520.
327 David N. Livingstone, "Geography, Tradition and the Scientific Revolution: An Interpretative Essay," *Transactions of the Institutes of British Geographers. New Series* 15, no. 3 (1990): 365.
328. Ibid. See also C. Glacken, Traces on the Rhodian Shore: Nature and Culture in Western Thought from Ancient Times to the End of the Eighteenth Century (Berkeley: University of California Press, 1967).
329 Livingstone, "Geography, Tradition and the Scientific Revolution," 366. See also David N. Livingstone, *The Preadamite Theory and the Marriage of Science and Religion* (Philadelphia: American Philosophical Society, 1992).

> Whether depicting the evolution of the world's landforms, tracing the historical spread of Christian civilisation, itemising the boundaries of nations, scrutinising regional diversities of people and places, or mediating on the mysteries of the hydrological cycle, the physico-theologists believed that the investigation of the world's design led inexorably to belief in an omnipotent divine agent … an argument postulating that the belief in God could be rationally inferred from foundational propositions.[330]

Believers of the school of physico-theology sought out examples that they felt reinforced religious principles, in addition to supporting scientific knowledge. Divinity was associated with design, because both European and non-European architecture had their origins in sacred principles, and through the proliferation of religious built form as part of typical urban language, legitimized the coexistence of religion and this new science.[331]

Experience over Authority

The sixteenth and seventeenth centuries prioritized "experience over authority," and principles of evaluation were introduced to assist in presenting what was considered a "rational" perspective.[332] Principles such as "objectivity, consensus, inductivism, experimentation, empiricism"[333] replaced traditional understandings and substituted them with an empirical perspective. Tradition dictated views based on observation, faith, and revelation, together with reasoned thought. By contrast, empirical knowledge resulted from experimentation and critical observation, rejecting all of the premises of traditional thought. At the same time as science became an important field in the cultural makeup of society in the sixteenth and seventeenth centuries, it did not directly resume the status of modern-day scientific thinking within the community. Acclaimed scientists such

---

330 Livingstone, "Geography, Tradition and the Scientific Revolution," 366.
331 Ibid., 366–67.
332 Ibid., 360.
333 Ibid. See also B. Vickers, *Occult and Scientific Mentalities in the Renaissance* (Cambridge: Cambridge University Press, 1984); C. Webster, *From Paracelsus to Newton: Magic and the Making of Modern Science* (Cambridge: Cambridge University Press, 1982).

as Johannes Kepler, Tycho Brahe, and Francis Bacon were also fascinated by astronomy, and Brahe actually discovered a new star. Brahe's astrological readings predicted that this star would be the cause of political upheaval in northern Europe in the sixteenth century (refer Figure 9).[334] Astrological beliefs were not abandoned in favor of science but rather a repositioning occurred, instead of a rejection of either scientific or astrological thinking. Such examples demonstrate that there was not a straightforward transition from premodern scientific thought to modern scientific principles.

Geography also followed the path of science and transitioned from premodern modes of thinking to modernity. In the early stages of this cycle, geography was linked to cosmography, as can be seen in, for example, William Cunningham's *The Cosmographical Glasse*, which was published in 1559, and navigation and hydrography were also considered within in its subject field. The main intention was to introduce readers to concepts of latitude, longitude, triangulation, and surveying, as well as terminology related to these practices. In addition, however, it recalled social science relationships put forward by ethnographers, therefore tying attributes of cultural behaviors to a place and context.[335]

Similarly, in the late sixteenth century, John Dee, a mathematician and scholar, as well as a mapmaker and a manufacturer of instruments, made notable contributions in the field of geography. Unfortunately, during his lifetime, his hypotheses were typically scorned because of his beliefs in occult practices.[336] Through his study and subsequent translation of Euclid's *Elements and Geometrie*, he defined geography from the perspective of location and as a subject that encompassed the totality of a situation of a town, city, or village.[337] Another

---

[334] Livingstone, "Geography, Tradition and the Scientific Revolution," 361.
[335] Ibid. See also David N. Livingstone, "Science, Magic, and Religion: A Contextual Reassessment of Geography in the Sixteenth and Seventeenth Centuries," *History of Science* 26 (1988): 269–94.
[336] Livingstone, "Geography, Tradition and the Scientific Revolution," 362; I. Seymour, "The Political Magic of John Dee," *History Today* 39, No. 1 (1989): 29–35; Denis Cosgrove, "Geography and the Mathematical Arts: Considerations on Humanism, the Occult and Geography in the Late Renaissance" (paper presented at the Annual Conference of the Institute of British Geographers, Leeds, January 8, 1985).
[337] Livingstone, "Geography, Tradition and the Scientific Revolution," 362. See also N. H. Clulee, *John Dee's Natural Philosophy: Between Science and Religion* (London: Routledge, 1988).

discipline that was intimately connected with this notion of geography was chorography, or the study of a specific place. Dee's ideas of geography were connected through mathematics to astrology, numerology, and celestial relationships. Therefore, his theses of human affairs being interrelated with celestial phenomena bridged the microcosm of the human world and the macrocosm of the greater cosmos.[338] So, in effect, as the discussion of the above two scientists demonstrates, the emergence of modern scientific thought grappled with intellectual theorizations of the human and cosmic interfaces. It was a very problematic path of rationalization required of scientists, to make any sense of it all. Therefore, Columbus's expeditions to the Antipodes and firsthand geographic recounting proved earlier views put forward by Ptolemy's theorizations of the same places to be invalid on some accounts. Columbus, by challenging classical narratives of geography, is often seen as a precursor to the age of rationalism of the Enlightenment.[339]

Emergence of the Survey

Geography was the core science in new ways of seeing the world, and the function of the map facilitated this new understanding through representation. The survey in its process of rationalization aided the delivery of a particular view of the world. Developments in instrumentation contributed to this rationalism in attempting to portray a highly empirical type of mapping, with the *truths* of certain projections established because of the logic of mathematical calculations, and instruments that were deemed to read a reality.

Many surveying instruments came out of medieval innovations and were further refined for Enlightenment practices. The compass, however, has earlier origins: the Chinese were credited with its invention in the third century BC. Around the eleventh century, the Chinese utilized orientations with the south-pointing needle. In earlier centuries, the north point had prevailed, yet the Chinese used the southerly direction when their Mongol invaders conquered them. Once Muslim travelers encountered this technology, courtesy of the Chinese, possibly in the mid-thirteenth century, they also adopted the south point as the prevailing

---

338 Livingstone, "Geography, Tradition and the Scientific Revolution," 362.

339 Ibid., 364–65. See also A. Pagden, "'The Impact of the New World on the Old': The History of an Idea," *Renaissance and Modern Studies* 30, No. 1 (1986): 1–11.

orientation. In turn, the Muslims passed the concepts of the compass to the Europeans, and they assumed the northerly aspect as their priority, although examples exist of maps commissioned and prepared by Europeans in the fifteenth century that indicated south at the top, as well as Christian maps still typically indicating east at the top.[340] The lodestone appeared to swing to either direction just as easily, that is, north or south,[341] whereas, for portolan charts, it has been argued that the compass with north at the top facilitated the chart maker's determination of rhumb lines.[342]

During the Middle Ages and Early Renaissance, technology was available to obtain measured distances in surveying by both direct and indirect methods. For example, in the layout of buildings on a site or determining property boundaries, a system of premeasured rods and cords was employed. These rods or knots in the cords were the units of the measurements. For greater distances, larger sized measuring implements would be utilized. There was a large element of error because the rods or cords would stretch, sag, and so on, in the process of measuring. Especially for larger distances, the customary method would be to pace out the distance or, if it was too great a space to carry out this procedure, to employ a method of riding or walking at a fixed rate.

---

[340] Gordon, "Sacred Directions," 223. See also Li Shu-hua, "The South-pointing Carriage and the Mariner's Compass," *Tsing Hua Journal of Chinese Studies* 1, no. 1 (1956): 64.

[341] Lodestone (also loadstone) literally means way-stone, from the use of the magnet in guiding mariners. It is a piece of rock of pure magnetite, with magnetic qualities that would naturally align with the poles, either north or south, and was used as a primitive compass. J. A. Simpson and E. S. C. Weiner, *Oxford English Dictionary*, in *Compact Oxford English Dictionary*, second ed., s.v. "Lodestone

[342] Ibid., second ed., s.v. "Portolan." A portolan is a book of sailing directions, describing harbours, sea coastlines, etc. Charts within this log also illustrate this information hence, the portolan chart reference Rhumb is "the line followed by a vessel sailing on one course. Any one of the set of lines drawn on through a point on a map or a chart and indicating the course of an object moving always in the same direction." Simpson and Weiner, *Oxford English Dictionary*, second ed., s.v. "Rhumb."

The shadow scales and the alidade of the astrolabe [343] [344] were considered the most accurate tools to work out simple trigonometric calculations during this period.[345] If the heights of distant buildings were known, these instruments could establish the distance at the ground between the observer and the tall building. Although these measurement exercises were possible at this point in history, they were not commonly adopted. Treatises of the Early Renaissance suggest these distances were often of relevance in military campaigns when calculations were required to establish whether gunners could reach their targets.

Because accurate measuring was necessary for the production of survey maps for the cases of building or property claims, in the fifteenth century, architect and engineer L. B. Alberti detailed the construction of a hodometer in his mathematical treatise.[346] However, although Leonardo da Vinci studied the construction of both the hodometer and pedometer[347] in his early-sixteenth-

---

[343] "An astrolabe is an instrument formerly used to take altitudes and to solve other problems of practical astronomy." Its form varied but one form was a planisphere, representing the circle of the heavens in the plane of the equinoctial, with movable sights. Historically used by astronomers, navigators, and astrologers, its many uses include locating and predicting the positions of the Sun, Moon, planets, and stars, determining local time given local latitude and vice versa, surveying, triangulation, and casting horoscopes. It was used in classical antiquity, and through the Islamic Golden Age, the European Middle Ages and Renaissance for all these purposes. In the Islamic world, it was also used to calculate the Qibla and to find the times for Salah (prayers). Simpson and Weiner, *Oxford English Dictionary*,, second ed., s.v. "Astrolabe."

[344] "An alidade (archaic forms include alhidade, alhidad, alidad) is the revolving radius of a graduated circle. The index of an astrolabe, quadrant or other graduated instrument, carrying the sights or the telescope, and showing the degrees cut off on the arc of the instrument. In the astrolabe it revolved at the back and was called by Chaucer the rule." Ibid., second ed., s.v. "Alidade."

[345] Juergen Schulz, "Jacopo de' Barbari's View of Venice: Map Making, City Views, Moralised Geography Before the Year 1500," *The Art Bulletin* 60, no. 3 (1978): 434.

[346] Hodometer also refers to the surveyor's wheel and odometer; see the following definition: Measures the distance traversed by a wheeled vehicle, consisting of a clockwork arrangement attached to the wheel or bearing which measures the number of revolutions. Simpson and Weiner, *Oxford English Dictionary*, second ed., s.v. "Hodometer." See also "The odometer of Vitruvius was based on chariot wheels of 4 feet (1.2 m) diameter turning 400 times in one Roman mile (about 1400 m). For each revolution a pin on the axle engaged a 400 tooth cogwheel thus turning it one complete revolution per mile. This engaged another gear with holes along the circumference, where pebbles (*calculus*) were located, that were to drop one by one into a box. The distance travelled would thus be given simply by counting the number of pebbles." Wikipedia contributors, "Odometer," *Wikipedia*, accessed April 20, 2012, http://en.wikipedia.org/wiki/Odometer.

[347] A pedometer is an instrument for measuring the number of steps taken and thus approximately measuring the distance traveled on foot, usually somewhat resembling a watch in size and appearance, having a dial plate marked with numbers around which a pointer of index-hand travels. Simpson and Weiner, *Oxford English Dictionary*, second ed., s.v. "Pedometer."

century drawings, he actioned the measurement of distance with the tradition of the time, that is, pacing them out on foot.[348]

By the middle of the sixteenth century, surveyors were measuring distances with chains, which were premeasured and composed of links that were the units of measuring. A graduated circle was used to establish bearings for measuring larger distances. In conjunction with an alidade to determine the point to be located, the surveyor would look across the graduated circle, in relation to the reference mark, which had been drawn on the circle, to correspond to a landmark in the field.

> The number of units between the reference mark and the line of sight provided an angular value that could be exactly reproduced in a drawing. One laid the circle upon the paper, aligned its reference mark once more with the chosen landmark or direction as represented on the paper, and counted off the number of units observed in the field.[349]

This instrument in the seventeenth century was known as the "Holland Circle,"[350] and the principles behind its determining of bearings are employed in the theodolite[351] of modern day. Principles of triangulation, although not in common use until the eighteenth century to establish bearings on a plot, were understood by Alberti in his fifteenth-century treatise.[352] For its time, surveying was

---

[348] Schulz, "Jacopo de' Barbari's View of Venice," 435.

[349] Ibid., 435–36.

[350] "In the late 16th century, a Dutch scholar and instrument maker named Gemma Frisius suggested that a good surveying instrument could be made by attaching a magnetic compass to the back of an astrolabe. Building on this idea, Jan Dou designed an instrument with two pairs of fixed sights attached to a graduated circle, and another pair of sights attached to either end of a movable alidade. Dou published an account of this instrument in Dutch in 1612, and the form became popular with surveyors in the Netherlands. The name Holland Circle was coined in the 19th century. The Holland Circle is similar to but easily distinguishable from the common theodolite." National Museum of American History, accessed April 20, 2012, http://americanhistory.si.edu/surveying/type.cfm?typeid=12.

[351] A theodolite is a portable surveying instrument originally for measuring horizontal angles, and consisting essentially of a planisphere or horizontal graduated circular plate, with an alidade or index bearing sights. Simpson and Weiner, *Oxford English Dictionary*, second ed., s.v. "Theodolite."

[352] Schulz, "Jacopo de' Barbari's View of Venice," 436.

considered an appropriate and the most precise or scientific method available to determine geography as well as one's place in the world and one's relations to others.

4.2 Enlightenment and Geographical Realism

Europeans sought a new geography and one that conveyed a sense of realism that was transparent and legible to all. The new geography involved replacing the mythical for a representation of the perceived truths of geographical space. Leading into the Enlightenment, mapmakers and philosophers turned to past conceptions of the world, such as the ancient representations of Ptolemy, to review how to interpret and thus represent new ideas about space. Geography of the seventeenth and eighteenth centuries was one of realism and an exact science.

New Geography

Around 1570, another publication grappled with the changing nature of geography utilizing elements of cosmography to produce cartography. Flemish cartographer and humanist Abraham Ortelius produced *Theatrum Orbis Terrarum*, which presented geographical information with images accompanied by a textual supportive account. The information within the atlas resembles a systematic combination of cosmography, geography, and chorography. According to Cosgrove, *Theatrum* does not respond in detail to previous works, such as Sebastian Munster's *Cosmographia*, and his ideas around mathematical cosmography, but rather attempts the most current view of global geography via a "spatial narrative."[353] However, as an appendix to later publications of the *Theatrum*, Ortelius added a separate section devoted to "Sacred and Ancient Geography."[354]

This series of maps and imagery, which was known as the *Parergon*, clearly demonstrated Christian religious thinking in its references to the New Testament and its influence on medieval scholarship and cartographic interpretations of the world. Instructively, the *Parergon* illustrated a map of a

[353] Denis Cosgrove, "Globalism and Tolerance in Early Modern Geography," *Annals of the Association of American Geographers* 93, no. 4 (2003): 855–56.
[354] Ibid, 856. See also G. Mangani, "Abraham Ortelius and the Hermetic Meaning of the Cordiform Projection," *Imago Mundi* 50, no. 1 (1998): 59–83.

journey paralleled in one of Rome's literary epic tales, Virgil's *Aeneid*, which dramatizes the ascent of Aeneas, the founder of the Roman race.[355] Aeneas carries out his work amid both a cosmic and an earthly landscape, therefore suggesting the existence and connection between two kinds of reality—one related to the broader, global universe and the other at a local and more intimate level.

Near the end of the sixteenth century, a European tradition of reading the universe and the globe from a number of philosophical perspectives was well established. Scientific, literary, artistic, and spiritual interpretations of global space were typical. In addition, however, Cosgrove highlights that the individual's role and understanding within the complex world theater is related to the greater context of the cosmos.[356] Thus, for individual human beings to realize their destiny, they needed to appreciate their moral position in the world, and their relative insignificance in the context of the expanse of the cosmos. Most conceptual connections between cosmography and geography remained to some extent in the seventeenth century, but such connections were largely separated by the eighteenth century. A separation between notions of the divine and the sovereign or stately, together with a shift in Enlightenment mapping practices, ensured the beginning of our modern way of rationalizing and viewing the world.

The seventeenth and eighteenth centuries affected utilization of certain forms of cartographic representation over others, because readers of history and geography increasingly needed to relate time to place.

> Book readers sought a history informed by precise cartography and a cartography that was in turn historically accurate—by contrast, for example, to the many early-modern illustrative woodcuts that used generic images of "a city" to represent Paris, Nuremberg, Venice, etc., at any date.[357]

---

[355] Cosgrove, "Globalism and Tolerance," 859. See also R. Jenkyns, *Virgil's Experience: Nature and History: Times, Names and Places* (Oxford: Clarendon Press, Oxford University Press, 1998); P. Hardie, *Virgil's Aeneid: Cosmos and Imperium* (Oxford: Clarendon Press, Oxford University Press, 1986).
[356] Cosgrove, "Globalism and Tolerance," 862.
[357] Black, *Maps and History*, 8.

Drawing

Scientific developments in optics emphasized the value of being able to sight the world clearly. In painting and cartographical terms, this equated to relaying the world to the viewer with a sense of "geographical realism."[358] Drawing techniques such as the production of the modern two-dimensional plan or ichnographic view were preferred by the end of the sixteenth century.[359] It was not until the eighteenth century that it was adopted as a common form of cartographic representation. The city view, consisting of a three-dimensional skewed perspective drawing, permitted a relatively easily interpreted, albeit inaccurate, depiction of urban space. It was also considerably more legible to the generally uneducated public who might attempt to interpret a map of the city. In addition, the expense, time, and preparation involved to deliver an ichnographic plan limited its commissioning and subsequent mainstream distribution.[360]

In her study of eighteenth-century Italian cartography, Barbara Naddeo uses the example of the *Mappa Topografica* to highlight the shift in ichnographic language to convey specific information about the urban form of Naples. Unlike earlier city views of Naples, any ichnographic form of representation could not sustain a dialogue of the social fabric of the city, the urban units appeared to extend into the countryside's plots, and the imposing city walls were not distinctive above the overall urban planimetric profile. Therefore, an aspect of decline and reality of the physical limits of a great city were implied in the ichnographic representation. By comparison, the city view conveyed a story of grandeur, density and expanse of the city fabric, and history of the architecture and presence of the city as opposed to its limitations.[361]

---

[358] Ibid., 8.

[359] Ichnography is a term that was defined by Vitruvius (i.2) "as the ground plan of the work, i.e. the geometrical projection or horizontal section representing the plan of any building, taken at such a level as to show the outer walls, with the doorways, windows, fireplaces, etc., and the correct thickness of the walls; the position of piers, columns or pilasters, courtyards and other features which constitute the design, as to scale." Wikipedia contributors, "Ichnography," *Wikipedia*, accessed April 20, 2012.

[360] John A. Pinto, "Origins and Developments of the Ichnographic City Plan," *The Journal of the Society of Architectural Historians* 35, no. 1 (1976): 50. See also Cosgrove, "Globalism and Tolerance," 69.

[361] "The *Mappa topografica* not only recorded the total area encompassed by the structural islands of the city, but made plain that those islands were significant as units of occupied land; it connoted that the city could not accommodate further construction and by way of implication, was an exporter of future growth. The only alternative signs within the city were the floor plans of

Therefore, the Enlightenment's reasoned thought and observation highlighted the importance of collaboration between all aspects of society to convey a useful interpretation of context, and laid the foundation for our modern conception of the map as a universal tool.[362] By graphic implication and as a precursor to our modern plan view, the ichnographic form enabled the determination and display of certain functional relationships that could not be ascertained in the city view, such as urban expanse and limitations of a city. Eighteenth-century projections still differed from the modern map in that they persisted with "multiple representational strategies and spatial constructions to enumerate and describe the world."[363]

Mapping and Reconnaissance

The map metamorphosed in a century from an accepted yet ambiguous sociocultural construct to a passive representation measured by "exact science."[364] The "Second Great Age of Exploration" was for the virtue of scientific discovery instead of only military and mercantile pursuits.[365] Therefore, the eighteenth century was also the period when geographers became connected with maps via the mapmaking process. Two practices enabled the creation of the geographic archive: first, *reconnaissance*, and second, *mapping*. Reconnaissance reflected Europe's colonial desires whereby expeditions would result in detailed narrations

---

institutions, old and new, which had established Naples as a highlight of the Grand Tour and evidently deprived it of civic space. While the omnipresence of ecclesiastical floor plans made a visual argument for a paucity of secular space, those of loyal royal institutions accented the two-dimensional depiction of the city with an expansive centrifugal rhythm, situated as they were at the peripheral heads of the major axes of the eighteenth-century capital." Barbara Ann Naddeo, 2004. "Topographies of Difference: Cartography of the City of Naples 1627–1775," *Imago Mundi* 56, Part 1 (2004): 38.

362 Matthew H. Edney, "Reconsidering Enlightenment Geography and Map Making: Reconnaissance, Mapping, Archive," in *Geography and Enlightenment*, eds. Charles W. J. Withers and David N. Livingstone, 165–198 (Chicago: University of Chicago Press, 1999), 190. Unfortunately, however, probably because of the sheer volume of material, "the conceptual unity of that [Enlightenment] archive was undermined by the unavoidable messiness of the archive's actual implementation. The archive contained a wide variety of materials housed in many different physical locations." The process of translation of collated information was an inherent problem in the Enlightenment, as it still is today.

363 Ibid., 193.

364 Suárez, *Early Mapping*, 27.

365 Edney, "Reconsidering Enlightenment Geography," 167–68. See also William H. Goetzmann, *New Lands, New Men: America and the Second Great Age of Discovery* (New York: Viking, 1986).

of places and routes. By the eighteenth century, these narrative accounts were quite different to earlier versions of descriptive geography. The act of reconnaissance during the Enlightenment was carried out by a variety of individuals of varying degrees of expertise, including merchants, engineers, surveyors, and scientists.

The art of reconnaissance was crucial to the age of colonial exploration and expansion. Although reconnaissance maps ranged in their graphic and discursive presentation, they were a valuable form of archive for colonial powers. Formats included narrated directions, in the form of distances and lists of dot point directions to essays of descriptive text capturing the geographers' presence within a visually overwhelming landscape. Both textual accounts and graphic imagery, from sketches to cartographic measurements, contributed to the overall construction of the recounted space and environs. The method by which information was selected for inclusion in the discussion of the spatial environment was dictated by the geographer's reasoned judgment and evaluation of worth. Unlike modern geographical accounts, Enlightenment reconnaissance was not particularly analytical in its discussion of space, but primarily descriptive.

By contrast, the process of mapping may be considered the opposite to that of reconnaissance. The reigning definition of a map in the 1700s was "the drawing together and presentation of the relations and hierarchies between phenomena. The space in which these phenomena are related was initially geographical."[366] However, by the middle of the century, a shift had occurred: mapping was no longer necessarily related, or related solely, to geographical space; maps were connected to taxonomies of knowledge, for example, "the mapping of nature."[367] Therefore, maps of the Enlightenment were primarily regarded as a means of ordering knowledge. These mapping constructs, contained information negotiated from the accumulated data of reconnaissance that were then displayed in a graphic format. As an example, Edney recounts that Tobias Mayer's *mappa critica Germaniae* of 1753 was intended

---

[366] Edney, "Reconsidering Enlightenment Geography," 186. See also Matthew Edney, "Mathematical Cosmography and the Social Ideology of British Cartography 1780–1820," *Imago Mundi* 46, no. 1 (1994): 101–16.

[367] Edney, "Reconsidering Enlightenment Geography," 186.

> to advertise the lack of firm, reliable geographical locations in central Europe and to demonstrate the resultant variability of geographical outlines. Of the two hundred towns in his map, there were observed latitudes for only 33 and observed longitudes for none. Like all other geographers, Mayer had to define the positions of towns from reconnaissance accounts, either by geodetic calculations or by plotting routes between known points in the graticule.[368]

To aid understandings of past processes of mappings, geographers of the eighteenth century recounted in their memoirs methods of teasing out geographic knowledge from other peripheral information. These memoirs served as documents legitimizing a geographer's life work, assumptions, and reasoning behind inclusion of material; therefore, they would often accompany the resultant maps.[369]

Technology developed further in the late eighteenth and early nineteenth century and shifted to accommodate a greater accuracy of processes of surveying such as triangulation. Triangulation also aided the process of observation in its reconciling the landscape with mapped space to a far greater degree than was achievable with the graticule. Through the organization of a mass of control points across the landscape to be surveyed, it was possible to achieve far larger and more accurate surveys than observational latitude and longitude were ever able to do.[370] Although the degree of accuracy achievable with triangulation was obvious in the late eighteenth century, because of the cost and laborious nature of implementing the technology, European countries did not really embrace such technologies until the mid-nineteenth century.

---

368 Ibid. Graticule is "a transparent plate or cell bearing a grid, cross-wire or scale, designed to be used with an optical instrument … for the purpose of positioning, measuring, or counting objects in the field of view." Simpson and Weiner, *Oxford English Dictionary*,second edition, s.v., "Graticule."

369 Edney, "Reconsidering Enlightenment Geography," 187. See also Mary Sponberg Pedley, "'Commode, complet, uniforme, et suivi': Problems in Atlas Editing in Enlightenment France," in *Editing Early and Historical Atlases: Papers Given at the Twenty-ninth Annual Conference on Editorial Problems*, ed. Joan Winearl, Toronto, University of Toronto Press, 1995.

370 Edney, "Reconsidering Enlightenment Geography," 191.

4.3 Where was Paradise? Sacred Geography Remapped

In his study "Geography, Enlightenment, and the Paradise Question," Charles Withers draws a comparison between "early modern botanic gardens," which were perceived as extensions of the Garden of Eden, and heavenly paradise of the Christian faith.[371] The botanic gardens were considered important reference points for early modern science, and their association with gardens of paradise reflected the Enlightenment predicament and apparent conflict between sacred understandings and reasoned thought.

> To the churchman, if not the exegetical philosopher, paradise was textual truth, a spiritual state, and a celestial place. Belief in the first at least demanded a geography for paradise.[372]

For premodern Christians, notions of earthly and geographical space were inseparable from ideas about paradisiacal space. During the Enlightenment, such understandings transferred to cartography, establishing a conflict between imaginative and literal ideas about space, and geographical representations of these concepts. For Christians to accept paradise as a place, Withers argues, there had to be a geographical understanding or image linked to paradise, and the posing of the question "Where was Paradise?"[373] A number of options existed for the response to this question, the first being that paradise had been lost and, therefore, the fall of a well-established Christian culture to a less significant existence had occurred. The next possibility for consideration was that the Flood had forsaken paradise and, therefore, through a process of geographical inquiry, exploration, and discovery, its whereabouts would be revealed. However, a prevailing view of the early medieval period depicted in *mappaemundi* representations was that paradise lay to the east. The four rivers the Tigris, Euphrates, Pison, and Gihon were typically indicated in these maps as originating from paradise. However, such representations created confusion when the

---

[371] Charles W. J. Withers, "Geography, Enlightenment, and the Paradise Question," in *Geography and Enlightenment*, eds. David N. Livingstone and Charles W. J. Withers, 67–92 (Chicago: University of Chicago, 1999), 67.
[372] Ibid.
[373] Ibid., 69.

positioning of the real Tigris, Euphrates, Ganges, and Nile Rivers could not be determined geographically.

On Columbus's 1498 expedition, he heard from his men that they had seen four rivers at the top of the Gulf of Paria, and they equated that with the description of the rivers of paradise.[374] Until the advent of technological improvements in surveying and mapping with what is known as the geographical Age of Discovery, it was not possible to prove empirically the existence of paradise. The promise of attaining such evidence of paradise to verify long-standing beliefs of the Christian world motivated many expeditions, in particular, those of the Portuguese.

Therefore, another belief in paradise resonated for explorers in the seventeenth century traveling to the New World because qualities of paradise were no longer imagined for an established Europe. Utopian ideas were transferred to new territories for European powers, and "pastoral paradise" replaced the "imperfections of western civilization."[375] Mapping provided an opportunity to explore these differing ideas and sites of paradise. During the sixteenth century, the Bible displayed maps of paradise or Eden in celestial and cosmographical forms.[376] Post-Reformation, the center of paradise was considered the city of Jerusalem because it was the symbolic center of the Christian world.[377] Yet different mapmaking European centers portrayed earthly paradise in a number of different geographical locations. Catalan maps of the fifteenth and sixteenth centuries indicate anywhere between East Africa and India as the virtuous location, and the Portuguese identified South America as a possible siting. By the seventeenth century, degrees of skepticism had arisen in the cartographic community as to an actual geographic location on earth, as theories of the location of the Garden of Eden become increasingly more elusive to empirical proof. Representations of paradise shifted in the mid-seventeenth century from being

---

[374] Ibid., 70. See also John Prest, *The Garden of Eden: The Botanic Garden and the Re-creation of Paradise* (New Haven: Yale University Press, 1981).
[375] Withers, "Geography, Enlightenment, and the Paradise Question," 70. See also Plaut, "Where is Paradise?"
[376] Withers, "Geography, Enlightenment, and the Paradise Question," 71
[377] Ibid. See also Delano-Smith and Morley Ingram, *Maps in Bibles 1500–1600.*

part of cartographic material to being decorative imagery and series of vignettes on the borders of maps.[378]

During the Enlightenment, the view of paradise changed from being a possible geographical discovery to being a phenomenon that had been wiped out by the Flood, implying a failure on the part of explorers to uncover its geographical location. Despite a lack of geographical evidence, the view persisted that an earthly paradise existed, and did not quash enthusiasm for other methods of discovery to be employed. Nathanael Carpenter, in his *Geography Delineated Forth in Two Bookes* (1625), proposed a combination of empirical, divine readings, and faith. He also referenced theories put forward by his contemporaries and recounted ancient belief systems to verify paradise on earth. From an empirical viewpoint, it was impossible for paradise to be located on the top of a mountain because it would impede the path of light from the sun, it would also be too obvious to the "eyes and knowledge of men," and lastly, it would appear more significant or "over-poize" earth. In conclusion, Carpenter asserted that paradise had been lost somewhere in Mesopotamia given that there was no other conceivable place where all of the sacred rivers would meet.[379]

The alternative view argued in Thomas Burnet's account "Sacred Theory of Earth," published in 1684, proposed the whole of earth as paradise. This series of four books has been equated with some of the most influential early Enlightenment theories on the relationship between ideas behind origins of natural history and tensions of biblical belief systems.[380] Burnet stated his method as follows:

> We may in the mean time observe … how preposterously they go to work, that sets themselves immediately to find out some pleasant place on earth to fix *paradise* in, before they have considered, or laid any grounds, to explain the general conditions of it, wheresoever it was. These must be known and determined, and we must take our aim and directions from these, how to

---

378 Withers, "Geography, Enlightenment, and the Paradise Question," 72; James Duncan, *Milton's Earthly Paradise: A Historical Study of Eden* (Minneapolis: University of Minnesota Press, 1972).
379 Withers, "Geography, Enlightenment, and the Paradise Question," 73.
380 Ibid., 75.

> proceed further in our enquiries after it; otherwise we sail without a compass, or seek a port and know not which way it lies.[381]

Therefore, Burnet's work is based on the sacred texts being true, and his work understanding the origins of nature set about explaining those so-called truths. Only when scriptures could not explain his findings were secular authors consulted to validate his ideas further. The search for earthly paradise continued into the mid-eighteenth century for many authors, although even the clergy acknowledged the inabilities of empirical geography to determine an actual site of existence. "Many learned Men of ancient times, very able in other respects," Bishop Huet wrote, "have been very ignorant of Geography … The old geography is not very certain."[382]

As Withers points out, these works identified views that were changing from concern about a present sacred geography to the search for a past geography of paradise. Therefore, there was no enduring concern for the failure of geography to determine a paradisiacal place, but rather that exploration had determined, by observation and account, that the Garden of Eden did not exist on earth, which was therefore a triumph for geography.[383] By the eighteenth century, quite another question emerged with regard to concerns of paradise, and that was about the type of language that was spoken. Therefore, the notion of paradise endured, despite the lack of geographical certainty about its physical location, as the subject of inquiry regarding its sociocultural characteristics, which by default suggested its existence. Fields of geographical inquiry, together with the history of language, culture, and evolution of peoples of different races, became a focus of this period of the Enlightenment. This expansion of scientific thought to bridge cultures and countries prompted thinking about ordering knowledge in a universal way for shared understandings of the findings. Structuring knowledge via taxonomies or universal signs and symbols enabled a common scientific language to evolve and

---

381 Ibid., 75–76; Robin A. Butlin, "Ideological Contexts and the Reconstruction of Biblical Landscapes in the Seventeenth and Early Eighteenth Centuries: Dr Edward Wells and the Historical of Geography of the Holy Land," in *Ideology and Landscape in Historical Perspective*, eds. Alan R. H. Baker and Gideon Biger (Cambridge: Cambridge University Press, 1992).
382 Ibid., "Geography, Enlightenment, and the Paradise Question," 77.
383 Ibid; T. Comito, "Renaissance Gardens and the Discovery of Paradise," *Journal of the History of Ideas* 32, no. 4 (1971): 483–506.

be interpreted by peoples who conversed in languages of different linguistic origins.[384] Bridging cultures and countries through graphic communication systems was topical, given the colonial cartographic undertakings of the Enlightenment.

4.4 Mapping and Colonial Desires

In their desire for control, the colonial administration remapped indigenous territories and, in effect, recreated the geographies of colonial spaces. The cartographical enterprise that resulted from the Second Great Age of Exploration possessed new territories in which to implement the most current Enlightenment principles. The structuring of knowledge through taxonomies was one such principle, and it was readily embraced to establish a legible and comprehensive archive of data ranging from indigenous plants and foods to the ethnic diversity of the colonies. Heinrich Berghaus's map shows one example of an attempt to reveal a European-centered view of the ethnic divisions of race. The information associated with this map also related to such factors as diet, population density, and life expectancy.[385] Therefore, the map may be seen as a visual taxonomy, as well as conveying geographic space.

Archiving and Bureaucracy

The French excelled at the colonial enterprise of preparing an archive of their colonial territories. The British, Dutch, Spanish, and Portuguese were all composing the data of their colonial conquests. The French appeared to have the most comprehensive bureaucratic machine for constructing their Enlightenment archive. The French entered the colonizing role wholesale in the seventeenth century. One example of their meticulous archiving and bureaucratic agenda was in the area of medical services, crucial to both French ex-patriots and colonial citizens surviving exposure to new foreign sicknesses. To combat wholesale outbreaks of disease in the eighteenth century, the French government prioritized training naval doctors for service in the colonies. Research into the health of its

[384] Withers, "Geography, Enlightenment, and the Paradise Question," 78.
[385] Whitfield, "Image of the World," 120–21.

citizens at home and abroad motivated experimentation with cooking with seawater (such as when making bread) and preserving fresh water while at sea.[386]

However, controlling and maintaining colonial space was more than merely maintaining a vigilant stance over the health of its citizens. Surveillance and control of the seas surrounding the colonies were prioritized; in addition, close attention was paid to developments in the sciences such as astronomy, cartography, and geodesy to gain greater insight into all aspects of colonial space. This approach prompted the establishment of collaborations between institutions such as the Académie des sciences, the Observatoire royal (1667), and the Académie de marine (1752).[387] The Académie des sciences improved upon determining longitude at sea with the development in the 1760s of John Harrison's marine chronometer.[388] Generous prize money sponsored by the Académie in its program of 1714 bequeathed money to marine and surveying research. Prior to this innovation, sailors at sea used astronomical observations and calculations to determine their position. Yet cartography remained a keen preoccupation of the French and, like other colonial powers, they took the surveying of colonial space and the building of cartographical information very seriously. In 1720, the French created the Dépôt des cartes et plans within in its naval ministry to compile an archive of hydrographical and mapped information.[389] An engineering corps as well as army and naval surveyors performed surveying duties in the colonies.

In addition, detailed taxonomies were prepared of the natural phenomena of the colonies, such as scientific notes of the flora and fauna native to each of the localities. Missionaries often provided such data from the field in collaboration with indigenous monastic activities.[390] As the colonies became more established,

---

[386] James E. McClellan III and Francois Regourd, "The Colonial Machine: French Science and Colonization in the Ancien Regime," in "Nature and Empire: Science and the Colonial Enterprise," ed. R. MacLeod, *Osiris*, 2nd Series, 15 (2000): 34.
[387] Ibid., 36. See also Josef W. Konvitz, *Cartography in France, 1660–1848: Science, Engineering and Statecraft* (Chicago: University of Chicago Press, 1987).
[388] The marine chronometer: "An instrument for measuring time … They differ from watches in having a more perfect escapement and a compensation balance and are used for measuring longitude at sea, and for other exact observation." Simpson and Weiner, *Oxford English Dictionary*, second ed., s.v., "Marine chronometer."
[389] McClellan and Regourd, "The Colonial Machine," 39. See also Dava Sobel and H. Andrewes, *The Illustrated Longitude* (New York: Walker, 1998).
[390] McClellan and Regourd, "The Colonial Machine," 41.

this tracking and recording of natural history shifted toward a process of economical production of plants, which would be useful for food and other purposes and thus signaled the introduction of exotic species to colonial regions. Spices such as cinnamon, nutmeg, pepper, and clove were introduced to the West Indies after raids on the Dutch East Indies and the Philippines were conducted by the French in the 1750s. Coffee trees were also introduced to the French West Indies, transforming the production of this commodity—a strategy that has resonated even into modernity.[391] The motivation of French authorities to organize a bureaucratic force behind colonial endeavors was clearly a response to the mercantile and economic benefits of colonial expansion. The purpose of mapping was to represent the Enlightenment archive and to communicate the European model of society. Mapping claimed to chart objectively, and on a semi-scientific basis (in the case of thematic maps), information such as the five degrees of civilization proposed by Clark in 1822, "savage, barbarous, half-civilized, civilized, enlightened."[392] The reality of these maps shows the complete cultural subjectivity of such propositions, and demonstrates that Enlightenment mappings also moved to communicate non-geographical ideas.

Imaginal Remapped

Imaginative and mythical mappings of geography specifically associated with premodern peoples, which were an essential component of indigenous understandings of land and settlement, were erased in the new mapping enterprise undertaken by colonial powers. Again, all colonizers participated in these processes, but it was particularly evident in Britain's surveying of India. Essentially, the production of the British colonial archive was achieved by "the conquest, rule and administration" of the subcontinent of India.[393] According to Kalpagam, collation of the colonial archive was undertaken through a series of investigative operations, which were referred to as modalities:

---

[391] Ibid., 42. See also James E. McClellan III, *Colonialism and Science: Saint Domingue in the Old Regime* (Baltimore: Johns Hopkins University Press, 1992).
[392] Whitfield, "Image of the World," 116.
[393] U. Kalpagam, "Cartography in Colonial India," *Economic and Political Weekly* 30, no. 30 (1995): PE 87.

> These modalities were the historiographic, the observational/travel, the survey modality, the enumerative, the museological, the surveillance and the sanitary modalities.[394]

These modalities undertaken during the colonial period, in particular, those that achieved cartographic outcomes, contributed to modern representations of India. In the case of India, it was an enormous task to survey and map such a large land mass, and during the colonial period, mapping operations occurred gradually as new states came under British control. Administration from the mother country demanded the fixing and demarcation of boundaries to be as accurate as possible, especially for the raising of revenue through land taxes, as well as the identification of extents of jurisdiction of political powers.[395]

The first modern map of India was prepared by a French geographer in 1751–52, and it was accompanied by the detailed descriptions of the process in *Mémoire de M. d'Anville*. The map was based on travel routes and writings as well as coastal charts that were rudimentary. Once the British achieved colonial rule over Indian territories after the Battle of Plassey in 1757, surveying commenced on a dramatic scale in the conquered lands.[396] Surveying was often accompanied by the memoirs of the surveyor to highlight the decision making as well as difficulties negotiated in arriving at the survey data. Information ranging from each state's military capabilities to opportunities for revenue collection was included in memoir rationalizations. These studies and notes far surpassed any geographical requirements.[397] Memoirs might describe and account for geographic features, positions of places, the family histories of ruling classes, relationships between neighboring ruling states, the history of conflicts, and any

---

[394] Ibid., PE 87. See also Bernard S. Cohn, "The Anthropology of a Colonial State and Its Forms of Knowledge" (paper presented at the Wenner-Gren Foundation International Symposium No. 107, Tensions of Empire: Colonial Control and Visions of Rule, Mijas Spain, November 5–13, 1988).

[395] Kalpagam, "Cartography in Colonial India," PE 87.

[396] Ibid.

[397] Ibid., PE 88. Nicholas Dirks, "Colonial Histories and Native Informants: Biography of an Archive," in *Orientalism and the Postcolonial Predicament: Perspective on South Asia*, eds. Carol Breckenridge and Peter Van der Veer (Philadelphia: University of Pennsylvania Press, 1993).

existing assets of the state and rulers, but were not always limited to these points.[398]

Possibly the most contentious form of surveying was the revenue survey because it brought to the forefront different understandings of space between the British and indigenous peoples. Alexander Read instigated one path of alleviating this struggle. Sent to conduct revenue settlement after the British defeated the Tippu Sultan in 1792, he employed locals to undertake the surveying tasks. On some occasions, accruing information was prohibitively difficult in that Read and his men would be locked out of a village because mistrust of the colonial power infiltrated many regions.[399]

Another essay that seeks to understand the colonial project is Alex Tickell's "Negotiating the Landscape: Travel, Transaction and the Mapping of Colonial India," which states intentions to review

> and foreground ways in which the colonial map can be read as the product of cross-cultural negotiation or translation, rather than an unproblematic, descriptive tracing … in particular … the way these texts betray the ideological instability of cartography at the same time as they contribute to what Gayatri Spivak has termed the "far-flung" and heterogeneous project to constitute the colonial subject as Other.[400]

Therefore, the production of colonized space and the resultant cartographic representations reflect the layers of conflict and negotiation that produced that space. The Great Trigonometrical Survey, which was an exercise in cartographic representation as well as documentary texts, comprehensively reflected Enlightenment ideologies. Mapping this colonial space of India also brought to the fore the domination of the British as a political power. Through the mapping process, in its collection and navigation of information to put into the map, it also "neutralised the threat of its cultural and physical difference and, in

---

[398] Kalpagam, "Cartography in Colonial India," PE 88.
[399] Ibid., PE 93.
[400] Alex Tickell, "Negotiating the Landscape: Travel, Transaction and the Mapping of Colonial India," *The Yearbook of English Studies* 34 (2004): 20.

turn, presented the authoritative 'position of the [European colonial] subject as fixed and unchangeable.'"[401] Therefore, Everest, the surveyor, was obliged and pressured to collect the "right" information and ensure its accuracy as well as convey the appropriate political message. The survey thus represented the relationships that were negotiated to achieve its construction, that is, between all contributors ranging from the colonizers to the colonized, rather than any topographical study it claimed to demonstrate.[402] Everest's journal account of his progress across Bengal indicated his frustration with attempting to achieve elevated points in the landscape as points of bearing. To a certain degree, Everest's assumption was that he would be measuring an uninscribed landscape but, in actuality, Indian temples dotted the Bengal region as well as other parts of India, and often they were the most elevated points. Unlike his counterpart Lambert, who utilized these religious high points to bear off, Everest constructed his own towers at great expense so as not to impose upon the local religious structures. These colonial towers became entwined in local traditions, either through being disassembled as unsavory marks on native territory or by according the towers with magic and inclusion in local native ceremonies. Therefore, an imposed spatiality of colonization began to enmesh with the local spatiality.[403] In practice, these surveying projects involved large teams of people, and some of the party was always ahead of the main group, which created problems of passing information between groups. Leader groups may have faced hostile locals and then some subversion of information may have taken place, or simply by the time the remainder of the group caught up to the leading party, the data may have been revisited. Any numbers of discrepancies were possible, given the length of time these surveys took to achieve. The daunting task of covering so much ground and the negotiation of hostile villagers were complicating elements affecting the outcome of the survey.

Because of the complex nature of the task, translating landscape elements into an intelligible survey was inevitably fraught for all parties involved in the

---

[401] Ibid.

[402] Ibid., 22.

[403] Ibid., 24. See also, on colonial space, and agency of the colonized, Brenda S. A. Yeoh, *Contesting Space: Power Relations and the Urban Built Environment in Colonial Singapore* (New York: Oxford University Press, 1996).

mapping process. For example, the locals had certain associations with some landscape elements of intrinsic meaning, whereas to the British surveyors their significance was as a landscape feature or a mark on a map. The surveyor Francis Buchanan reported his problems of cultural translation in Bengal:

> The geographers of Europe are apt to be enraged, when in tracing a river they find that an inconsiderable stream falling into the grand channel changes its name, and that the source of this smaller stream is obstinately considered by the natives as the source of the river, either having been the first to which they had access, or having at one time been the largest.[404]

No doubt, this cross-cultural predicament occurred frequently. The locals would have been unwilling to concede generations of historical verbal accounts as incorrect, and the colonizers would have required a preciseness or accuracy from the locals' knowledge but at the same time might have often cast doubt on any colonial aspersions.[405] In a less assertive manner, but which may also be interpreted as the reconfiguring of the colonial space, Everest records instances in his journal when locals treated the surveyors' interruption into their social space as a spiritual experience. Surveying instruments and practices of the colonizers were considered by some of the local infirmed community as an opportunity to engage with objects of "miraculous powers."[406] Tickell suggests such acts by the native locals may be considered a recasting of the trigonometric points within the site as another cartographic reading, such as an opportunity for a spiritual pilgrimage, which also forces a reevaluation of the intended cartographic representation.[407]

---

[404] Edney, Mapping an Empire, 331.
[405] Tickell, "Negotiating the Landscape," 26.
[406] Ibid., 27.
[407] Ibid. See also Setha M. Low and Denise Lawrence-Zúñiga, *The Anthropology of Space and Place: Locating Culture* (Malden, MA: Blackwell, 2003.

4.5 Concluding Remarks

From medieval conceptions, a new vision of geographic space from its slow transition to a new conceptual vision of the world, from the sixteenth century to the nineteenth century, highlights critical changes in Western cartography and, eventually, non-Western geography. A recasting of geographic understanding to an empirical conception of the world meant these ramifications were expressed and transformated in maps. Geographic science was integral in reshaping all conceptions of space from socioreligious to humanist views. The recasting of sociocultural and traditional spatial understandings was the predominant procedure when occupying colonial territories. An exposition of this colonial remapping enterprise follows in the next chapter.

CHAPTER FIVE
Remapping the World

The Enlightenment meant that modern science desired and constructed a more literal interpretation of geography. The modes of production of mapping also equated to a much narrower use of the map, as well as a very small spectrum within which maps were normally read. This new way of seeing and conceptualizing the world, from early modernity to the Enlightenment, prompted an enormous investment in the Western cartographical enterprise, from developments in instrumentation to exploration and colonization. These technologies and practices enabled a revisiting of mapping with the endeavor of producing accurate and precise representations of space on maps. Motivations for an accurate geography on a global scale ranged from providing reliable information for sea passage in exploration to mapping territory for imperial expansion. At a domestic level, empirical and quantitative mappings were perceived necessary for land ownership, revenue raising, infrastructure purposes, judicial disputes, and generally, the ordering of settlement. These desires were largely a preoccupation of the West, and slowly manifested into Southeast Asian societies around the nineteenth century. A remapping of the world occurred because the Enlightenment signified a completely new way of seeing and thinking about geographic space.

5.1 Mapping and the Scientific Gaze

Enlightenment mapping processes were defined by their structuring of space and their representation of observation.[408] Generally, these maps were widely regarded as:

> a dialectic of experience and reason: it is experience, usually visual in nature, which produces information about the world, but it is reason which refines and structures that experience and gives it meaning.[409]

Despite the emphasis on attaining empirically derived mappings, in actuality, the Enlightenment approach always produced "highly idealized" representations in the "reasoned manipulation of data."[410]

Eighteenth-century geographical knowledge was not limited to the map, but included observations about the physical world that constituted the mapped space. These factors ranged from the material, such as architecture and sociocultural form, to environmental characteristics, such as landforms and climate.[411] However, this information was charted in a two-dimensional format, rather than represented in the pictorial method demonstrated in the seventeenth-century projection of the bird's-eye view of Amsterdam. The geographical process in the seventeenth and eighteenth centuries was also one of comparison, whereby an existing mapped region would be analyzed against a newly mapped region. The chart by Edmund Halley (1702) is an example drawn because of comparisons, determined by his astronomical expertise.

---

[408] Tickell, "Negotiating the Landscape," 27. See also Pedley, "Commode, complet, uniforme, et suivi".

[409] Edney, "Reconsidering Enlightenment Geography," 175. See also Paul Laxton, "The Geodetic and Topographical Evaluation of English County Maps, 1740–1840," *The Cartographic Journal* 13, no. 1 (1976): 37–54.

[410] Edney, *Mapping an Empire*, 190. See also Andrew S. Cook, 1985. "More by Accident than Design: The Development of Topographical Mapping in India in the Nineteenth Century" (paper presented at the Eleventh International Conference on the History of Cartography, Ottawa, July, 1985).

[411] Edney, *Mapping an Empire*, 42. See also Geoffrey C. Gunn, *First Globalization: The Eurasian Exchange, 1500–1800*, World Social Change, ed. Mark Selden (Lanham, MD, USA: Rowman & Littlefield, 2003).

According to Barbara Stafford, observations of comparison were conducted from two different intentions.[412] The "scientific" gaze compared with the "picturesque" gaze dealt with different perspectives, and thus had different objectives for the collation and representing of data.[413] The Enlightenment period was said to be characterized by the blending of aesthetics with scientific reasoning, each of which were "embedded in its own cultural constructs."[414] European mapping practice was dominated in the eighteenth century by mapping undertaken by reconnaissance (experience and geographical narrative) as well as by locating objects in space through the empiricism of latitude and longitude. Therefore, Enlightenment mapping was considered, for its time, the most precise scientific method available to determine one's place in the world and one's relations to others. Zürner's map is a comprehensive example of this shift to a modern way of thinking and illustrates the transition of mapping from an artistic concept of the seventeenth century to a "newly conceived image fitting the Age of Enlightenment."[415]

Science

The scientific gaze involved observing what was perceived as the facts and individual characteristics of the landscape in detail. Among the scientific innovations that influenced conceptions of mapping were visual and mechanistic ways of seeing. The *camera obscura* technology of the mid-1600s was embraced as the main method of recording images of the world.

Jonathan Crary, in his analysis of the role of the *camera obscura* in "Techniques of the Observer," suggests that the viewer becomes abstracted from the scene viewed.[416] However, "linear perspective" through the lens of the camera was legitimized as the main premise for visual records.[417]

---

412 Barbara Maria Stafford, Voyage into Substance: Art, Science and the Illustrated Travel Account, 1760–1840 (Cambridge, MA: MIT Press, 1984), 5.
413 Edney, *Mapping an Empire*, 54.
414 Gunn, First Globalization, 140. See also Goetzmann, New Lands, New Men.
415 Whitfield, "Image of the World," 106–107.
416 Ron Broglio, *Technologies of the Picturesque: British Art, Poetry, and Instruments, 1750–1830* (Lewisburg, PA: Bucknell University Press, 2008), 58.
417 Edney, *Mapping an Empire*, 46–47.

> Of all the senses, sight was the most mechanistic, least sensuous, and hence was closest to the truth. Sight's almost exclusive privilege as the means by which to know the world was only reinforced by the use of artificial technologies of vision. In addition to the *camera obscura* itself, telescopes and microscopes underscored the mechanical character of vision by enhancing human faculties and so allowing more of creation to be seen and examined.[418]

Sight could also be further quantified with the use of measuring equipment such as graduated rulers, and less tangible components of the environment could be measured, such as temperature with the advent of thermometers. Enlightenment observation attempted to reveal the "truths" of the scene, by "observing rules, codes, conventions and practices. Observation is guided vision; the observer looks at the world in a controlled manner."[419] The premise by which vision could be considered a reliable method for accounting for geographical facts came about through the second aspect associated with Enlightenment thinking, and that was using the power of reasoning. Reason enabled observers to make sense of things and hinged on the capacity of the human brain to evaluate the circumstances and contributing factors of objects of study. Therefore, the limiting aspect for individuals in developing these thinking processes was the extent of their education. Typically, the model for Enlightenment thinking was developed through Western education systems.

An understanding of rational thinking during the Enlightenment was based on various theorizations, including Bernard de Fontenelle's 1699 *ésprit géometrique* and editors Denis Diderot and Jean d'Alembert's *Encyclopédie* (1751–65), which proposed a process of observation whereby objects were compared and all of their individual characteristics evaluated and measured against one another. In addition, taxonomies developed by botanists such as

---

[418] Ibid., 48. See also Jonathan Crary, *Techniques of the Observer: On Mission and Modernity in the Nineteenth Century* (Cambridge, MA: MIT Press, 1990).
[419] Edney, *Mapping an Empire*, 48. See also Margarita Bowen, *Empiricism and Geographical Thought: From Francis Bacon to Alexander Humboldt* (Cambridge: Cambridge University Press, 1981).

Linnaeus (Karl von Linné) served as examples of how to organize and group geographical data.[420] Further influences came from the field of natural history and the development of systems of evaluations that would delimit observations. The method undertaken to observe these systems was through comparison and flexibility, while being guided by the objects' individual attributes and character. By the eighteenth century, another practice of systemizing had emerged, whereby a broader spectrum of difference was permitted, observed, and recorded. Projects for classification would then extend beyond "vegetable, animal, and mineral worlds into the realms, for example, of analytical geometry, mechanics, linguistics and natural society."[421]

The Enlightenment's collective archive was connected to conceptions of cartography, as a result of the evaluations of objects that needed to be cited within a physical space or location. This space then needed to be identified relative to other geographical spaces or locations in order to revisit it or measure the validity of assumptions about it. Cartographic metaphors featured in Linnaeus's taxonomic systems and Michel Adanson's method for accruing knowledge. Maps were recognized for assisting the ordering of observed material to create knowledge.[422] Measuring apparatuses had evolved substantially by the late eighteenth century, and enabled mathematical modeling of many observations from the field. Therefore, post-1750, the sciences of cartography and geography had come to rely on the empiricism associated with the recording of data by instrumentation of all varieties. However, the premise—observation achieved the truths of the observed, which subsequently were rationalized through processes of comparison and a further reconciling of those observations—endured as a credible evaluation process throughout the Enlightenment. One of the main reasons for its endurance was that the process of reasoning was considered sufficiently flexible to explain any anomalies of observation.[423]

---

420 Edney, *Mapping an Empire*, 49. See also Robert Home, *Of Planting and Planning: The Making of British Colonial Cities* (London: Spon, 1997).

421 Edney, *Mapping an Empire*, 50. See also Michel Foucault, *The Order of Things: An Archaeology of the Human Sciences* (New York: Random House, 1970).

422 Edney, *Mapping an Empire*, 51. Richard Yeo, "Reading Encyclopedias: Science and the Organisation of Knowledge in British Dictionaries of Arts and Sciences, 1730–1850," *Isis* 82, no. 1 (1991): 24–49.

423 Edney, *Mapping an Empire*, 51.

The crucial consideration for recording the geographical archive in the eighteenth and nineteenth centuries was the accurate depiction of location. Defining particular locations on a map was achieved through the act of positioning—derived from longitude and latitude calculations as reference points—and subsequently recording these data on a map.

> The use of instruments to measure the positions of the planets and the fixed stars and so derive geographical locations exemplified the period's empiricism in that it entailed the observation of quite abstract phenomenon to derive concrete, quantified facts.[424]

Latitude was recorded by observing the position of the sun at midday above the horizon or its altitude—the angular elevation of the sun above the horizon. This latitude measurement was achieved with calculations that recognized the position relative to the celestial pole. By comparison, longitude measurements were somewhat arbitrary because no equivalent meridian to the celestial pole was apparent. A surveyor would measure longitude based on a universally agreed zero meridian. Yet this interpretation opened the measurement of longitude to large errors. Longitudinal difference was related to time, and time differences related to a three-hundred-and-sixty-degree equatorial circle. Therefore, fifteen degrees equated to a twenty-four-hour time variation. In the seventeenth century, Galileo proposed the moons of Jupiter, given their regular and reliable observed movements, as the basis for the construction of time. By the eighteenth century, in Paris and Greenwich, there were published schedules of the predicted eclipses of Jupiter's moons. This meant that surveyors in the field could determine their longitudinal measurements either side of these observations.[425] For maritime navigation, it was difficult to determine the position of Jupiter's moons because of the heaving and lurching of the ship, so in the mid-1700s,

[424] Ibid., 85. See also Mary Louise Pratt, *Imperial Eyes: Travel Writing and Transculturation* (London: Routledge, 1992).
[425] Edney, *Mapping an Empire*, 87. See also Stafford, *Voyage into Substance*.

longitude became determined in marine navigation via the lunar cycle, which was more regular than Jupiter's eclipses.[426]

The route survey remained the most popular form of mapping from the seventeenth into the nineteenth century because it was fast and required only basic instruments. The accuracy and detail of the survey depended upon the skills and the equipment of the traveler. However, the simplest route survey revealed a geographical narrative accompanied by the time to travel between certain landmarks or villages, and therefore an account of distance relative to direction of the sun. Direction was measured with a compass, and a perambulator measured distance, or the time of travel was measured by noting a specific speed. As a linear mapping of the landscape, these surveys were inherently inaccurate, because of the inaccuracies presented from compass readings and because the travel-time distance was very subjective. In addition, the perambulator did not keep a straight line, and its driver missed holes, crevices, and other impassable objects.

By the eighteenth century, trigonometrical surveys had become popular in aiding the surveying of large regional areas. Route surveys and linear calculations were viewed as too inaccurate to be used as legitimate methods of representing large areas; therefore, triangulation was embraced to gauge heights and distances, applied with a triangulated grid laid across the survey field. The grid was used in conjunction with astronomical observations at one or two points within this triangulated field. This process was used to locate the survey frame on the greater earth's surface. Initially conceived for the field of geodesy and the measurement of the earth's size and shape, triangulation depended on the knowledge of certain benchmarks. The measurement of the size of the earth was crucial to Enlightenment's rational understanding of the world. Therefore, attaining this

[426] Edney, *Mapping an Empire*, 87. In addition, the invention of the chronometer (see note 388) in 1760 meant a device that would keep fairly accurate time for long-haul sea voyages. A navigator could then calculate longitude based on Greenwich time relative to the position of local stars and the sun. Overall, however, chronometers were limited in their general use because of their fragility for land travel and were too expensive for mass production and use in general surveying. Therefore, the most common method, which endured well into the nineteenth century, for the determination of longitude, was the observation of the eclipse of Jupiter's moons. The lunar tables prepared by Nevil Maskelyne were an adaptation of Tobias Mayer's (astronomer royal) 1754 lunar tables that enabled the "angular parallax in the moon's position as seen from Greenwich and from a ship would give the longitudinal difference."

knowledge was at the most sophisticated end of science during the eighteenth and nineteenth centuries. This structured and ordered method of viewing the world through triangulation meant that spatial characteristics were now being related to one another. Therefore, errors were supposed to be more easily distinguished with a greater field of reference points to cross-correlate information and provide checkpoints.[427]

Aesthetics

In comparison with the scientific gaze, the picturesque gaze represented a more socially acceptable manipulation of landscapes viewed. In the 1800s, the picturesque movement was interpreted as a combination of the themes of "Roman pastoral poetry with the style of Italianate landscape art."[428] Poetry formed an essential element of a proper Englishman's education, and Italian landscape paintings were understood from aristocratic travels around Europe. The arrangement of picturesque spaces reflected stage-like scenes, with the foreground often in shadow as if it was being slightly obscured by the wings of the stage, and shadows cast by these features. Therefore, the middle distance in the picture was often the focus and would display a city or an important village building such as a church or castle. The background was also obscured because of the light cast by the horizon. The intent of these compositions as idealized constructs was to evoke emotive responses from their viewers. Tourist maps and guidebooks enforced the existence of the picturesque in their recounting of landscapes where one might witness such emotive imagery. Of course, as idealized views of landscape, these views could not be recaptured. Yet there were instructions in guidebooks as to how one might achieve the same view or experience of the picturesque. They might list details of the place to grasp the colors and perspectives seen in popular paintings of the period.[429]

Ann Bermingham suggests that people in picturesque images are absorbed into the landscape because they appear oblivious of the beauty of the environment

[427] Edney, *Mapping an Empire*, 108. See also Andrew S. Cook, 1985. "Alexander Dalrymple and John Arnold: Chronometers and the Representation of Longitude on East India Company Charts," *Vistas in Astronomy* 28, Part 1 (1985): 189–95.

[428] Edney, *Mapping an Empire*, 57. Also Cook, "Alexander Dalrymple and John Arnold."

[429] Edney, *Mapping an Empire*, 58.

around them.[430] By contrast, our appreciation of the surrounding beauty in its textures and colors highlights our different spatial and visual experience of the subjects. In the same way, this image illustrates a contrived and romantic view of the landscape, and this tradition transplanted understandings of the picturesque to the New World via European officers and government officials. They measured their view of these landscapes through eyes that were used to viewing landscape in a particular way. The images and creative representations were then sent back home, perpetuating biased and idealized views of colonized landscapes. These practices were not necessarily noted as criticisms, but rather to highlight that bias and subjectivity were considered typical and critical components of human assessment and evaluation.[431]

In his studies, Paul Carter identifies the crossover between geographical, travel, and explorative narratives.[432] Carter asserts that it was not possible to write a geographical narrative without being influenced by other literary accounts of the period. Geographical records attempted a mechanical approach to observations, yet were presented in text and maintained distance between the observer and the object under observation.[433] Unlike the explorers' view of the world, the focus of geographical accounts was not discoveries, but everyday occurrences and, to an extent, the less interesting aspects of the landscape. Inherent ambiguities were associated with the social and personal aspects to viewing the landscape because all observations are influenced by social practices and social beliefs. Therefore, through observation and representations such as maps, the landscape acts as a social construct within these parameters.[434] The similarity may be drawn with the modern map, in that it exists as a social construct whereby the landscape may be seen from a particular point of view.

Stafford and others have identified the picturesque gaze and the scientific gaze as opposing one another; they also served the purpose of providing different

---

430 Ann Bermingham, *Landscape and Ideology: The English Rustic Tradition, 1740–1860* Berkeley: University of California Press, 1986), 139. See also Hugh Prince, "Art and Agrarian Change, 1710–1815," in Cosgrove and Daniels, *Iconography of Landscape*.

431 See Said, *Orientalism*, 159. Said preferred to consider such subjectivity as an intended and conscious power struggle because, he argued, "[t]o be a European in the Orient, *always* involves being a consciousness set apart from, and unequal with, its surroundings."

432 Carter, *Road to Botany Bay*, 65.

433 Edney, *Mapping an Empire*, 66.

434 Ibid., 72.

perspectives and an archive of information with which the geographical gaze could engage. The two genres of observation gave the geographer an overview and a spatial context to investigate, which in turn facilitated the use of mapping. Mapping acted as the process within which the overview could be represented and where notions of spatial concepts were played out. Individual observations within the spatial construct enabled geographers to locate objects within the frame of mapped space. They did not need to create a map themselves. By observing an object, locating it within a spatial frame, becoming aware of its relations to other objects, and then moving through the landscape, relations to the broader context resulted in a kind of mental mapping.[435] The geographer produced an archive of work that included both observation and narrative.

5.2 Exotic Lands, Colonial Encounters

The allure of the spice trade was one aspect that prompted European interest in the Southeast Asian region, but the "Age of Discovery" was really a symptom of complex intellectual, technical, and political forces in play at the time.[436] European fifteenth- and sixteenth-century conceptions of what we now understand as modern Southeast Asia entailed exotic lands known as "the Indies." (Refer Figure 10).[437] This understanding derived from a Hellenistic account from antiquity of all the lands east of the Indus River. Ptolemy and Strabo revised such images with their maps, which were in turn replaced with geographical information supplied by the Portuguese. The naming of these lands remained as determined in the Ptolemaic map, *India extra gangem*, or the Indies outside of the Ganges, or all lands west, which included Indochina and modern Indonesia.[438]

As is evident from the part map of Java by Gerard van Keulen, conceptions of the world had shifted dramatically in two hundred years (refer Figure 11). Vignettes display characteristics of built form as well as the geography, rather than a preoccupation with the native, in addition to a greater knowledge of topography and extents of the island.

---

[435] Ibid., 64.
[436] Whitfield, "Image of the World," 36.
[437] Norman G. Owen, *The Emergence of Modern Southeast Asia* (Honolulu: University of Hawai'i Press, 2005), 3. See also Donald K. Emmerson, 1984. "'Southeast Asia': What's in a Name?" *Journal of Southeast Asian Studies* 15, no. 1 (1984): 1–21.
[438] Owen, *Emergence of Modern Southeast Asia*, 3.

According to Norman Owen in *The Emergence of Modern Southeast Asia*, the eighteenth century denoted a significant shift in sociocultural patterns of the life of local peoples of the region.[439] Southeast Asians had always been affected by natural characteristics of their climate, geography, river and ocean systems, rainforests, and alluvial plains. Yet the eighteenth century was characterized by changes of geographical, territorial, and spatial understandings while also subject to the undermining of many of the Southeast Asian natural resources, effecting changed sociocultural relationships, political practices, and modes of production of settlement.

Geography

Exotic lands were deemed to have exotic and valuable spices, which traders from diverse cultural backgrounds sought as early as the thirteenth century. A key aspect of the Southeast Asian geography was its location relative to other trading partners by sea. The Malacca Strait enabled a strategic port for buying spices but also for breaking the long sea journeys from Europe. Geographic realities of Southeast Asia have always played a significant role in everyday life for its inhabitants and their realization of settlement. The large annual rainfall in this tropical region provided optimal conditions for harvests of rice, palm oil, rubber, coffee, tea, and pepper, among other crops that also featured as strong exports (as well as being utilized for local consumption).[440] Aspects of climate and environment also affected locations of settlement and choices of transportation to and within Southeast Asia.[441] In earlier days of the spice trade in the Malacca Strait, the monsoon winds ensured ships would dock at this port. It was not until the winds had shifted that ships could resume their route of passage for trading of spices and textiles. Again, once steam power prevailed in the nineteenth century, modern shipping was no longer affected by the limitations of the prevailing wind direction. These changes in technology also affected the regularity of ships docking in the Malacca Strait. Another dominant environmental characteristic of Southeast Asia is their impressive river system,

---

[439] Owen, *Emergence of Modern Southeast Asia.*, 5.
[440] Ibid., 6.
[441] Ibid., 9.

which enabled premodern societies such as the Khmers from Cambodia to harness the floodwaters of the monsoon for farming during the dry season. Rivers such as the Mekong, Ayeyarwady (Irrawaddy), and Thanlyin (Salween) in Myanmar, in addition to Chaophraya in Thailand and the Red River in Vietnam, supported food supplies and acted as vast transport corridors for the countries bordering them.[442] All of these geographical characteristics had implications for aspects of Southeast Asian urban environments, from their location to their morphological form, and were predominantly independent of mapping practice, until around the nineteenth century.[443]

Territory

Traditionally, Southeast Asian territories were conceived by indigenous peoples through geographical landmarks, which in turn defined political allegiance. For example, in the seventeenth century, the Vietnamese-Lao wars were mediated by the Vietnamese ruler and the Lao monarch through an agreement that "every inhabitant in the upper Mekong valley who lived in a house built on stilts owed allegiance to Laos, while those whose homes had earth floors owed allegiance to Vietnam."[444] Therefore, lines on a map of a defined space were superseded for "cultural preference," a traditional indigenous concept that no longer resonates for indigenous peoples of the twenty-first century.[445]

For the Europeans, geography and territory were intimately interwoven because determining territory extents necessitated boundary demarcation.[446] Europeans liked to identify boundaries in maps, to assimilate potential regions or countries to conquer. An understanding of the extent and limitations of geography was important, as was awareness of the peoples, their sites of habitation, sources of food, and sociocultural and religious practice. General knowledge as to whether the country or state was going to be a useful acquisition and the determination of a viable entry strategy were also important. In the context of

---

[442] Ibid., 8.
[443] Colonisation and mappings were adopted at different times in different areas of Southeast Asia.
[444] Owen, Emergence of Modern Southeast Asia, 10.
[445] Ibid., 11.
[446] Edney, *Mapping an Empire*, 1; Jorge Luis Borges, "Museum: On Rigor in Science," in *Dreamtigers*, eds. Mildred Boyer and Harold Morland (Austin: University of Texas Press, 1964).

contributing to the Enlightenment knowledge archive, graphically plotting this information gave another level of legibility and understanding.

In the case of the British in Southeast Asia, according to Owen:

> Colonialism had many unintended effects in Southeast Asia, as it damaged, destroyed, blended with, or was absorbed by local people and traditions in a complex, interactive process. In many cases, precolonial social practices (such as patron–client and family–centered politics) persisted through the colonial era and beyond. Others, like "strong-man" leadership, revived with independence, and still others, like tributary diplomacy, disappeared more or less for good.[447]

From a mapping perspective, the complete colonial region was fragmented into territories. There was great difficulty for the locals to comprehend why lands were separated in such ways and for what reasons. For example, Sumatra fell under the auspices of the Dutch and therefore there were imperial divisions on the Malay Peninsula, which was at that time dominated by the British. In contrast, Lao states had been closely aligned with the Siamese but were colonized by the French and thus governed similarly to Vietnam. Borneo was divided between the Dutch and the British. The western mainland became Burma and the British established its boundaries. The colonization process then fractured the sense of coherence that had existed in Southeast Asia between countries and meant that the colonies now looked to their mother countries for political and economic processes rather than maintaining support within the region. Overall, from a colonial perspective, emphasizing difference became a recognized program of colonial rule, in spatial organization of settlement as well as socio-politically within colonial society.[448] Ethnicities became the focus of colonial organization, indigenous and ethnic minorities were treated differently in terms of who were

---

[447] Owen, *Emergence of Modern Southeast Asia*, 78. See also Swati Chattopadhyay, "A Critical History of Architecture in a Post-Colonial World: A View from Indian History," *Architronic* 6, no. 1 (1997), cited September 27, 2004, http://architronic.saed.kent.edu/v6n1/v6n1.05a.html.

[448] Anderson, 1991, 78.

best suited and trusted with colonial tasks. Empowering some local groups over others created a hierarchical structure in society to serve colonial rule.

The motivation to take control of the entire region arose because Europe and the United States grew tired of dealing with the remaining ruling entrepôts.[449] Their colonizers drew up borders of Southeast Asian countries, and major intellectual and political processes occurred because of these actions. Traditionally, power had derived from religious and sacred sources, which then became tied to indigenous political authorities, but based on obligations and sacred blessings of spiritual leaders. In colonial Southeast Asia, new rulers were geographically remote and secular, and appeared to be arbitrary political figureheads. Government authority rested in economical achievement over intentions of gaining respect and status. Colonized peoples grappled with the paradox of holding onto their religious beliefs and sociocultural value systems derived from centuries of spiritual associations with their leaders, while adjusting and changing to what were completely new institutions of governance. The only exception in this region was the kingdom of Siam. Amid a rapidly shrinking physical sphere, the authority of the Siamese monarchy was able to hold onto its traditional form of governance and belief systems.[450] These moral, cultural, and geographical disparities between colonial and locals were highlighted in mappings because borders of colonial spaces were completely incongruous with indigenous conceptions (refer Figure 12).

Space

The map as a physical object is important as a representation of space. Mapmaking processes are also vital in understanding the limitations and imaginations that have been graphically conveyed.

> As with any other form of representation—graphic or textual, artifactual or ephemeral—meaning is invested in all aspects of cartography: in the instrumentation and technologies wielded by the geographer; in the social relations within which maps are made

---

[449] Owen, *Emergence of Modern Southeast Asia*, 201.
[450] Ibid., 203.

> and used; and, in the cultural expectations which define, and which are defined by, the mapped image.[451]

The British colonial approach for the Southeast Asian region was not nearly as organized nor comprehensive as it was in India. Nonetheless, confusion about indigenous conceptions of space, as well as the cultural diversity of the colonized, was a universal problem of the colonization process. In the case of James Rennell, surveyor-general to Bengal, his lack of understanding of the cultural diversity and socioreligious history of India resulted in him naming it Hindoostan. Through the colonial mapping process, local space was completely misrepresented. A large percentage of the population was Hindus, but Islamic invaders to the region had originally coined it Hindoostan. Rennell subdivided the country on his maps into Mughal sub-provinces, based on seventeenth-century distinctions achieved under the reign of Emperor Akbar, but at the same time, did not distinguish the East India Company territory of Bengal. Rennell's maps conveyed the Mughals as the predominant rulers over the whole subcontinent.[452]

> The geographical rhetoric of British India was so effective that India had become a real entity for both British imperialists and Indian nationalists alike. Both groups held "India" to be a single, coherent, self-referential geographical entity coincident with the bounds of the South Asian subcontinent and the extent of British power but which nonetheless predated British hegemony.[453]

This perception of British colonized space depended on the Enlightenment's philosophy of maps being designated as true and accurate representations of a geographic reality. Therefore, there was unquestioning belief in the information conveyed in Rennell's graphic depictions of India. The framing of India was possible, first, because of the view of Enlightenment knowledge and,

---

[451] Edney, *Mapping an Empire*, 2. Claude Nicolet, "Space, Geography, and Politics in the Early Roman Empire", in *Jerome Lectures* (Ann Arbor: University of Michigan Press, 1991).
[452] Owen, *Emergence of Modern Southeast Asia*, 11–12; Pauline Lunsingh Scheurleer, "Rich Remains from Social Anthropological Work in Eighteenth-Century India," *Journal of the History of Collections* 8, no. 1 (1996): 71–91.
[453] Edney, *Mapping an Empire*, 15.

second, because of what was seen at the time as a practical and scientific means of achieving a "true" reality of geography through methods of triangulation.[454] Technology culminated in the Great Trigonometrical Survey of India, which was believed by the British to be a comprehensive, accurate, ordered, and concise archive of geographical knowledge. Critical understandings of geography were possible via the vehicle of the map, owing to the expansion of mass production of printing processes. Both new print technologies and the rise of humanist and proto-capitalist culture could be argued as critical to the propagation of the map as an essential and integral tool of geography.

5.3 Cross-Cultural Mappings

The main difficulties of mappings between Europeans and indigenous communities were largely related to contrary views of regions or boundaries, different styles of representation, and purposes behind mappings.

Language

By the time of the Enlightenment, the scope of geographical analysis had expanded beyond the substantially narrow realms of "*geo*-graphy (earth-description) and *choro*-graphy (region-description)."[455] The judgments by which both the surveyor and the geographical observer took in the world were the eyes of mediation between what revealed itself and that, which was accounted and recorded. It was not necessarily how accurately spatial characteristics were measured, but *what* was measured that shaped the graphic representation or the map.[456] Another process that made a substantial contribution to our modern understanding of the archive of cartographical knowledge came about in the 1830s when William Morrison referred to a map's design and display of certain symbols and signs as consisting of a "language."[457]

By acknowledging a language of symbols, cartographic information could be rationalized and represented in a neatly comprehended fashion. All compiled

---

[454] Ibid., 17.
[455] Ibid.
[456] Ibid., 95.
[457] Ibid., 96. See also John K. Noyes, "The Representation of Spatial History," *Pretexts* 4, no. 2 (1993): 120–27.

data could be related through a framework that consisted of both latitude and longitude, laid within a format of graticules and meridians, and the remaining data situated within this frame. Information such as the translation of indigenous understandings of distance could be reduced to a standard European interpretation of units of distance, which could be universally applied in a very pragmatic way. Differences in conceptions of astronomical systems were evaluated and transferred to the limitations of the map. To explain the decision-making processes around the resultant graphic depictions, cartographers used their memoirs as extensive narratives on their numerous and inevitable rationalizations. The detail and construction of the analysis presented by the cartographer contributed to the perceived authenticity and accuracy of the cartographic construct.

Informants

The main recognition of indigenous groups was given to Filipino mapmakers, who were usually of Spanish origin but regarded themselves first as Filipinos. In addition, Malays have been credited with assisting in the mapping of Borneo and the Sulu Sea in the eighteenth century. Much of the information ascertained by Europeans included verbal accounts by indigenous groups of their experiences and knowledge passed down through the generations.[458] The degree of European mapmakers' exposure and firsthand experience of Southeast Asia and its peoples was quite diverse. Some mapmakers traveled extensively in Southeast Asia; others used secondary sources and data to produce their maps, never having traveled to the region. Notable names from the former group included Ferdinand Magellan and Francesco Pigafetta, who produced one of the earliest maps of the Spice Islands. Then, in the early seventeenth century, William Dampier charted many harbors, channels, and islands in the region. Others include Alexander Dalrymple, John Crawford, Jean-Marie Dayot, and James Horsburgh, all of whom lived and worked in Southeast Asia in mapping and charting capacities in the eighteenth and nineteenth centuries.[459]

---

[458] Fell, *Early Maps of South-East Asia*, 9.
[459] Ibid., 10–11.

Cross-cultural mappings occurred for possibly three hundred and fifty years after the arrival of the Portuguese in Southeast Asia. Typically, indigenous input in these cross-cultural encounters of cartographic processes was rarely recorded or attributed to their rightful owners.[460] Initially, great delays occurred in the transfer of geographical notes to the cartographers assembling the representations, generally because of the distance and time it took to reach Europe from Southeast Asia, as well as the high regard for secrecy to prevent other colonial powers seeking out new territories. In addition, the process of mapmaking at this time in Europe had become a labor-intensive task following the creation of new types of woodblocks and copperplates to depict more accurately new information about the Southeast Asian region.

Institutions such as the East India Company recorded their observations of the Southeast Asian landscape, and when this information found its way back to the mother country, it was interpreted literally. Accounts provided in a sequential fashion provided information about a range of topics from agricultural technologies to political observations. Translation of this information into legible graphic representations or textual accounts proved difficult, especially for those removed from the experience entirely.[461] Europeans on the ground often relied upon indigenous perspectives to further inform ambiguous accounts of spatial contexts, which typically entailed conflicting or confused observations by European observers.

Territories that were outside of foreign control, such as Siamese territories, relied almost entirely upon information from a broad and diverse number of indigenous sources. In the accounts of the British geographers accruing details on Siam, the local informants were typically introduced in such a way as to legitimate the surveyors' account. For example, surveyors might insist upon the level of accuracy of their instruments in the same way as they intended to justify the credible nature of their informant, who might be referred to as reliable, or in some way superior to the local person. Thereafter, any information was subjected to the principles of Enlightenment thinking whereby reasoning and corroborating

---

[460] Ibid., 9. See also John Pickles, *A History of Spaces: Cartographic Reason, Mapping and the Geo-Coded World* (London: Routledge, 2004).

[461] Edney, *Mapping an Empire*, 77. See also Michel de Certeau, *The Practice of Everyday Life*, trans. Steve Rendall (Berkeley: University of California Press, 1984).

of different accounts of the same object or place were evaluated. Informants were often not named unless they were significant members of the royal family or part of the local governing organizations. Evaluation of measurement in time also needed to be verified because concepts between locals and Europeans were often discrepant. Time was not a universal concept; for example, the Hindoo hour contained twenty-four minutes, so if an itinerary was recorded with a certain number of days to reach the next landmark, an adjustment of hours per day was required to rectify any inconsistencies.[462]

Another method for valuing the truth of data gathered by one informant compared with that gathered by another was to judge by the informants' reputations. When the informants' reputation was equal, the principle of reason would always be applied, the principle being that all information had an aspect of truth and, through comparison, the essence of truths would be revealed. This process suggests the degree to which compilation of data, observation of objects, and reasoning were intertwined. By the end of the day, the geographer would record his reasoned judgment of his and others' observations, which supposedly resulted in an evaluated and considered set of conclusions about the fieldwork. However, this process may also be evaluated as a more general portrayal of events, rather than a conclusive indication of the "actual" conditions under study. A distinct hierarchical structure established between the European and the local informant whereby, despite all of the checks and balances the colonizers put in place, there was never complete trust in information conveyed by local peoples. At the earliest opportunity, European surveyors would discard local information in favor of what their organizations considered more accurate and reliable sources, that is, of European origin. Underlying the European skepticism of local abilities to conceive of European-type mapped spaces were the natives' commitments to religious beliefs and cosmological concepts of space.

Survey

Europeans considered their geographical beliefs to be rational, measurable, plausible, and precise. Beliefs prevailed, such as those of Adam Smith in his history of astronomy:

---

[462] Edney, *Mapping an Empire*, 84.

> The reverence and gratitude, with which some of the appearances of nature inspire [the savage], convince him that they are the proper objects of reverence and gratitude, and therefore proceed from some intelligent beings … Hence the origin of Polytheism, and of that vulgar superstition which ascribes all the irregular events of nature to the favour or displeasure of intelligent, though invisible beings, to gods daemons, witches, genii, fairies.[463]

Trigonometry permitted the production of a more homogeneous frame for geographical knowledge. It also established a fixed field of operations within which errors could be resolved through mathematical means such as calculations. The process of triangulation enabled surveys to grapple with large amounts of information, even to the extent that large wall maps could not display every detail. Therefore, maps were divided into structures reflecting latitude and longitudinal coordinates. These divisions were purely pragmatic inscriptions laid across the landscape for "plane-table surveying."[464] The shift from surveys in the field to office-constructed surveys, the latter being the survey conceived through extensive mathematical computations, defined a new genre of mapping. Maps created through triangulation and trigonometric calculations reduced every spatial coordinate to an equivalence or uniformity. The surveyor was responsible for delivering the "truths" of the surveyed space, but the result was only a measured view of that reality. There is no other manipulation of reasoned or recorded peripheral data to achieve an understanding of space of the survey. This is the point where there is an epistemological shift in mapping processes and insight into how modernity's view of the world came about. Although more of the colony could be mapped and thus seen on paper, through the means of triangulation, it was only from a single viewpoint.[465]

---

[463] K. C. Cleaver, "Adam Smith on Astronomy," *History of Science* 27 no. 76 (1989): 212.
[464] Edney, *Mapping an Empire*, 113. See also Roger Hahn, "New Considerations on the Physical Sciences of the Enlightenment Era," *Studies on Voltaire and the Eighteenth Century* 264 (1989): 789–96.
[465] Edney, *Mapping an Empire*, 115.

The most inhibitive factor of the trigonometrical survey was the cost. The time required to conduct the fieldwork could be anywhere from weeks to months given the weather, visibility issues, and terrain-inhibiting sight lines. The list of interfering factors was extensive. As has been highlighted, difficulties existed in the shift of mapping practice from the field to the office and the successful relaying of information to achieve a comprehensive understanding of the area under survey. Many issues such as costs could have been overcome if the required resources had been available to the colonial administrators, but typically, funding proved inadequate.

Corpus

By comparison with European mappings of the region in the eighteenth century, the following indigenous maps continue to convey very different geographical conceptions. Only a very small number of remaining urban maps have been located by Joseph Schwartzberg in his studies of historical cartographies analyzing indigenous maps from Southeast Asia. These maps are deemed of a "cross-cultural" genre simply because there would have been knowledge of Europeans' cartographical practices. However, it is not obvious whether any cartographical traditions were intentionally embraced in these representations. The extent of the map corpus on this subject was found to relate only to Burma and was only sixteen in total.[466]

The "Malay Chart of the Malay Peninsula and the Gulf of Siam" the geography appears to be presented inaccurately, but the sequencing of places along the coast is correct with respect to inland landmark features. Inland areas do not reflect the same accuracies.[467] Schwartzberg believes this map was not intended to convey solely a river route but rather a route that traversed parts of the river and parts of the land. Part of the land route was over the hills of the Cameron Highlands. Although sea routes were the least expensive and therefore the preferred mode of transport if goods were required to come via land, given geographical limitations, such routes were also made possible for traders. Another motivation for pursuing alternative routes to the sea was the prevalence of pirates,

---

466 Schwartzberg, "Southeast Asian Geographical Maps," 796.

467 Phillimore, " Early Map of the Malay Peninsula."

particularly in the Malacca Strait. The purpose of this map is not entirely clear because it does not display or elaborate on sailing times and there is an absence of any directional guides. Effectively for the local seamen who traversed the waters on a regular basis, most of the information conveyed on this map could easily have been remembered without needing to consult the mapped representation.

By contrast, "The Map of Amarapura, Capital of Burma before the Founding of Mandalay," of mid-nineteenth century origins, demonstrates another style of mapping: plan forms of the city coupled with different projections of landscape elements (refer Figure 13). The map is a copy of the original, reputed to have been replicated by the monks of Taung Lay None Monastery in Amarapura in the 1970s.[468] The original mapping was in a four-panel format on parabaik (sheets of local papers joined and folded to form accordion-like books).[469] The map denotes both urban and non-urban features in an evocative style. An old royal walled city, monasteries and pagodas, the street layout, and the hamlets and wards outside the royal city walls provide a relatively comprehensive plotting of city and context. Unusually, compared with other Burmese maps of this era, this map is orientated toward the west. The royal city grid contrasts with the sprawling areas beyond, and extends to the mountain range in the background.

The intention behind the move of the capital from Amarapura to Mandalay was to give the Burmese citizens some reprieve from the crowded and dirty conditions of the former capital.[470] The mapmaker was not included among the surveyors and engineers instructed to lay out the capital. The construction of Mandalay began in 1857, and its mappings were said to have been completed in 1855.[471] No European has been referenced in connection with the construction of the capital, its design, or the mapmaking process, but Captain Fraser's work in Rangoon may have been an influence. If these dates are correct, it may be an example of projective mapping practices in Southeast Asia. Certainly, the Burmese king would not have been ignorant of the design and construction of Rangoon. There are a number of discrepancies between what exists on the ground

---

468 Schwartzberg, "Southeast Asian Geographical Maps," 798.
469 Ibid., 840.
470 U. Maung Maung Tin and Thomas Owen Morris, "Mindon Min's Development Plan for the Mandalay Area," *Journal of the Burma Research Society* 49, no. 1 (1966): 29, cited in Schwartzberg, "Cosmography in Southeast Asia," 798.
471 Ibid., 799.

at Mandalay and the undertakings as conveyed on the map. Spiritual references were considered of importance and a priority. The initial design of the building was in accordance with the plans; the perimeter of the royal compound formed a perfect square and, in Burmese measurements, equated to each side consisting of 600 *tas* (a Burmese yard). The total *tas* of the perimeter of the compound reached a distance of 2.06 kilometers or 2400 *tas*. This correlated with the number of years since the Buddha attained enlightenment.[472]

The Royal Temple Complex at the Foot of Mandalay Hill, which was possibly drawn around 1875, may also have preceded the temple construction. The buildings indicated in the plan include the Sandamuni Pagoda located between the royal garden and the interim palace grounds. The center of the plan includes the larger structure—the Kyauktawgyi pagoda and a library to the left of this central building, along with numerous rest houses indicated as small rectangles. The sacred Bodhi trees form the four points to mark the square in the larger pagoda. The upper left corner of the map indicates the location of the ordination hall or Pathana Sima.[473] These Burmese mappings suggest the beginning of a synergy between European and Southeast Asian mappings at a certain level. Yet Southeast Asian projections appear to hold on to some of their sociocultural belief systems, through their drawing style and emphasis upon religious forms.

5.4 Mapping Emerging Urbanity

Although Southeast Asia was typically considered to demonstrate low degrees of urbanization, Laura Lee Junker's research indicates that, compared with Europe during the same period, it was one of the most urbanized regions in precolonial times, given its low urban densities. Junker's premise for Southeast Asian urbanism is based on the assumption that urbanism may be understood as

---

[472] Schwartzberg, "Southeast Asian Geographical Maps," 799.

[473] Ibid., 801–802. The process for surveying for the Mandalay capital was proposed as follows: "A certain amount of rough triangulation must have been employed to fix the positions of the islands and the villages in the Irrawaddy, and of its right bank. Otherwise all the control seems to have been by measurements along and offsets from the lines of the grid. If all the grid lines shown on the plan were completely traversed to fix the positions of the villages, canals, rivers, lakes, and the mountain foot, it would mean that considerably more than 1600 miles of line were measured, recorded, and plotted during the survey."

"an index or ratio of concentrated 'city'- or 'town'-dwelling people versus the dispersed, rural component of the population."[474]

Precolonial

A census recorded in 1900 led the European community to believe that the Southeast Asian region demonstrated some of the lowest populated urban centers in the world. Anthony Reid concurs with Junker's view and believes the colonizing of Southeast Asia, from the seventeenth to mid-twentieth centuries, led to a period of de-urbanization for Southeast Asian cities and towns.[475] This effect was brought about by colonial powers excluding indigenous populations from the new European commercial centers, where once there was booming trade relations between indigenous and global markets. Prior to the seventeenth century, Reid accounts for several Southeast Asia cities being larger than their European trading partners; for example, Thang-Long (Vietnam), Ayutthaya (Thailand), Pegu (Burma), Melaka (Malaya Peninsula), Makassar (Maluku) and Mataram exceeded populations of between 100,000 and 200,000.[476] The structure of the precolonial inland Southeast Asian city often entailed a walled structure encompassing the dwellings of the royal family and related servants, with a relatively high urban density. The surrounding area was more dispersed and less tangible because boundaries were not clearly defined. Reid estimates that, in the sixteenth and early seventeenth centuries, over the Southeast Asian region, five percent of indigenous peoples were living in urban centers. This figure was lower than in Mughal India and Late Ming–Early Ch'ing China but larger than in the majority of European centers.[477] The urbanization index is the greatest for port cities such as Melaka,

[474] Laura Lee Junker, "Population Dynamics and Urbanism in Premodern Island Southeast Asia," in *Urbanism in the Preindustrial World: Cross-Cultural Approaches*, ed. Glenn R. Storey, 203–32 (Tuscaloosa, AL: The University of Alabama Press, 2006), 21.
[475] Anthony Reid, *Southeast Asia in the Early Modern Era* (Ithaca: Cornell University Press, 1993), 63.
[476] Anthony Reid, Southeast Asia, (1993, 68–73), cited in Junker, "Population Dynamics and Urbanism," 212.
[477] "In 1511, Melaka was a city of 100,000–200,000, with a spectacular series of mosques and a walled royal core (with buildings for royal elephants, artillery, and the reception of foreign dignitaries) ruled by Muslim kings." Armando Cortesão, *The 'Suma Oriental' of Tomé Pires: An Account of the East, from the Red Sea to China,* 2 vols., (1944) 1990., , quoted in Junker, "Population Dynamics and Urbanism," 212.

which catered for trade through the Malacca Strait to the Indian Ocean and South China Sea.

Maritime trade and the resultant economic boom-time conditions prompted the rapid growth of populations of these port cities. These growing populations were assisted by large influxes of foreign slave workers who carried out work ranging from farming to entertainment as concubines and domestic servants. In addition, given that, nobility in Southeast Asia was not so concerned with ties to landholdings but rather to proximity to trade centers, their alliances were with other wealthy patrons and aristocrats, who typically clustered around the trading cities.[478] These nobles built their elaborate dwellings and large gardens in the suburbs surrounding the urban center, which then gave the visual impression of urbanity on a large-scale to the early European observers.

Most architectural constructions of Southeast Asian cities were built out of perishable materials. Wood, bamboo, and rattan were among the most common materials for houses, which were often constructed off the ground, depending on the exact location within the Southeast Asian region. There were exceptions in Buddhist and Hindu cities, where the centrally planned core of the city contained stone and other durable structures, typically for religious purposes. Therefore, these cities such as Pegu and Ava in Burma and Siamese cities such as Ayutthaya have archaeological remains that have assisted urban historians in piecing together the original urban fabric of these premodern cities. Given that many domestic structures were not located on the ground, archaeologists have found it difficult to determine definitive settlement patterns, and thus spatial concentrations of villages between regions.[479]

Junker identifies Southeast Asian patterns of urbanization as consistent with what cultural geographers and anthropologists call a "dendritic" settlement system:

> Dendritic systems are characterized by the concentration of regional and political and economic control within a single primate center, which exerts weakening authority over a series of linearly

---

[478] Junker, "Population Dynamics and Urbanism," 215.
[479] Ibid., 218.

> radiating settlements … Most … occur in situations where long-distance trade plays a dominant role in the internal economy, concentrating economic advantage and political power with those settlements located favorably for … external trade systems.[480]

Constraints of geography, topography, and accessible transport routes such as rivers or access to coastal ports mean that there is a linear relationship between the transport route and the trade center or the core of urban development. This urban pattern results in a central wealthy, elite, and politically powerful urban center with a descending pattern of wealth as the density of settlement deteriorates out to the periphery of the urban center.[481]

To prevent traders going directly to the upriver villages or providers of the goods such as spices, horn, and other trade materials, alliances were formed between the rulers of the coastal cities and bureaucrats in these villages, which were strategically situated on a major river system. Jane Allen, in her study of hinterland trade in seventeenth- and eighteenth-century Kedah, identified archaeological evidence of upriver stations that were significantly more populated than the typical upland village. In addition, they were often fortified centers in contrast to neighboring villages.[482]

Compared with their neighbors such as China and India, historically, Southeast Asian urban development occurred quite differently. Many traditional Southeast Asian polities of organized administration and with an economic agenda typically became one city. For example, in Northern Java, premodern cities such as Tuban, Demak, and Gresik had their own political leaders as well as control of their economies. This was also true for Ayutthaya on the Chao Phraya River, which shared the sociopolitical practices established for the seaside towns of Jambi in Sumatra and Mandalay in Burma.[483]

---

[480] Ibid., 223.

[481] K. B. Kelley, "Dendritic Central-Place Systems and the Regional Organisation of Navajo Trading Posts," in *Regional Analysis: Vol. I Economic Systems*, ed. C. Smith, 219–54 (New York: Academic Press, 1976), 221, cited in Junker, "Population Dynamics and Urbanism," 223.

[482] Jane Allen, "Trade and Site Distribution in Early Historic Period Kedah: Geoarcheological, Historic and Locational Evidence," *Bulletin of the Indo-Pacific Prehistory Association* 10 (1991): 307–19, cited in Junker, "Population Dynamics and Urbanism," 226.

[483] Eric Tagliacozzo, "An Urban Ocean: Note on the Historical Evolution of Coastal Cities in Greater Southeast Asia," *Journal of Urban History* 33, no. 6 (2007): 915.

Colonial

Many of these urban patterns continued into the colonial period for Southeast Asia. As an example, for both the Spanish and American period, Manila continued to be the capital of the Philippines. As in the case of Jakarta for the length of the Dutch reign, Rangoon served as the main distribution center for the British and as a major political stronghold within the Southeast Asian region. These centers of politics and economics remained connected to indigenous outposts and responsive to local, regional, and global markets.[484] The endurance of these centers was a little uncertain in that they were entirely dependent on trade for their survival. In addition, they were intermediate ports along trade routes, rather than the beginning or the end-points.[485]

Other specific characteristics related to the innovative functioning of Southeast Asian commercial centers through their political and economic exchanges were in notions of center and periphery. The commercial center was important for encountering and realizing the trade, but typically, the hinterlands were responsible for producing the valuable items for exchange (refer Figure 14). The Dayaks, Bataks of Borneo, and Sumatra sourced hornbills, camphor, and gold from their jungle outposts; these goods passed through their commercial center but were typically bound for China. Ethnic minority groups in the lowlands of Siam and Burma grew medicinal foodstuffs and animal horns for trade, which facilitated complementary reliance between hinterland dwellers and the "middle-men" in the commercial centers. In turn, outlying regions gave political allegiance to the urban center, fostering the trade of their wares, despite the inaccessibility of the terrain to many villages, given the required economic ties with the urban centers. Through these mutual exchanges, both parties benefited. This in turn meant that outlying tribes based on the perceived need to maintain trade relations might agree demands of the central polity, such as conversion to a preferred religion.

Colonial intervention into Southeast Asian urban space through changing and increasing modes of production in manufacturing and infrastructure prompted wholesale migration to the cities. By the eighteenth century, two main urban

---

[484] Ibid.
[485] Ibid., 916.

forms dominated in Southeast Asia: coastal or port cities of the archipelago, such as Batavia, Manila, Bantam, Makassar, and Aceh. These port cities which sustained economically large commercial enterprises within their settlements, as well as other mainland cities participating in maritime trade, such as Pathein (Bassein) in Myanmar, Hoi An (Faifo) in central Vietnam, and Ayutthaya (Siam).[486] Siam was considered both a royal and port city. Royal cities dated back to precolonial Southeast Asian history and consisted of sites of ceremonial as well as administrative centers but were also bases for early manufacturing. Craftspeople were from both slave and free labor classes, and undertook services such as gold- and silversmithing, blacksmithing, leatherwork, and furniture manufacturing. There were also all kinds of domestic workers for the palace as well as teachers, dancers, official scribes in Arabic, Latin, or Indic scripts, advisors to the king, and so on.[487] In addition, priests, monks, scholars, astrologers, poets, and chroniclers were resident. Europeans, who also operated and owned the printing presses, governed these cities.

Anthony Reid states that the fifteenth to eighteenth centuries represented the "Age of Commerce" in the Southeast Asian region, and the nineteenth century reflected more the "Age of Peasantry," when wholesale agricultural enterprises predominated.[488] Wholesale production of the railway and road systems could be achieved given the existence of a large labor workforce. Steam power ensured that there was more surveillance of migration and workers, which translated to knowledge of borders on land and sea, which in turn legitimized territory on maps. This shift in the nineteenth century from transient populations to a new focus on fixing workers to a place and a state ensured different relationships between cultural groups throughout Southeast Asia. Traditional alliances to nation-states also shifted because inhabitants were not always indigenous to the locality but rather stationed in a place for making a living.[489] Therefore, when workers moved between nation-states they would be distanced from their traditional sociocultural practices because their life and work were focused on

---

[486] Owen, *Emergence of Modern Southeast Asia*, 33.
[487] Ibid.
[488] Anthony Reid, *Southeast Asia in the Early Modern Era*. (Ithaca: Cornell University Press,1993) cited in Owen, *Emergence of Modern Southeast Asia*, 33.
[489] Owen, *Emergence of Modern Southeast Asia*, 33–34.

achieving a livelihood. As a new way of residing became necessary for foreign workers, this signified that transitioning to a modern, rational style of living had emerged. Religious and cultural requirements adapted and perhaps became secondary to socioeconomic priorities.

In Ricardo Padron's essay "Mapping Plus Ultra: Cartography, Space, and Hispanic Modernity", he reiterates the thoughts of Henri Lefebvre, that Western production of space in its geometric and representational conception has been at the expense of the perceived or lived spatial experience.[490] Through mapping urbanity,

> traditional "representational spaces"—spaces as they are perceived—such as the hearth of the geography of the sacred, are correspondingly stripped of their authority. In this order of abstraction everything comes to be understood as either a location or an object within this space, and therefore becomes amenable to systematic understanding, commodification, appropriation, or subordination by a viewing subject.[491]

5.5 Concluding Remarks

In summary, the Enlightenment project foregrounded the rationalist thinking that was entrenched in mapmaking and other sciences by the nineteenth century. Colonial intervention into Southeast Asian indigenous space ensured a transitioning from socioreligious and culturally focused settlement forms to hybrid systems that embraced new geographies and concepts of space. A sense of aesthetic and functional "sameness" of urban environments seemed to emerge; this was the result of a number of factors, but the altered nature of mapping contributed to this condition.

---

[490] Ricardo Padron, "Mapping Plus Ultra: Cartography, Space, and Hispanic Modernity," *Representations* 79 (Summer 2002): 28.
[491] Ibid.

Figure 1: Circa 1086 World Map from *Ibn al-Wardi's Kharidat al-Ajaib wa-Faridat al-Gharaib* (The Unbored Pearl of Wonders and the Precious Gem of Marvels)
Illuminated Manuscript, 1481, 7.75 x 5.7 IN.
By permission and courtesy of the Library of Congress, Washington, D.C.

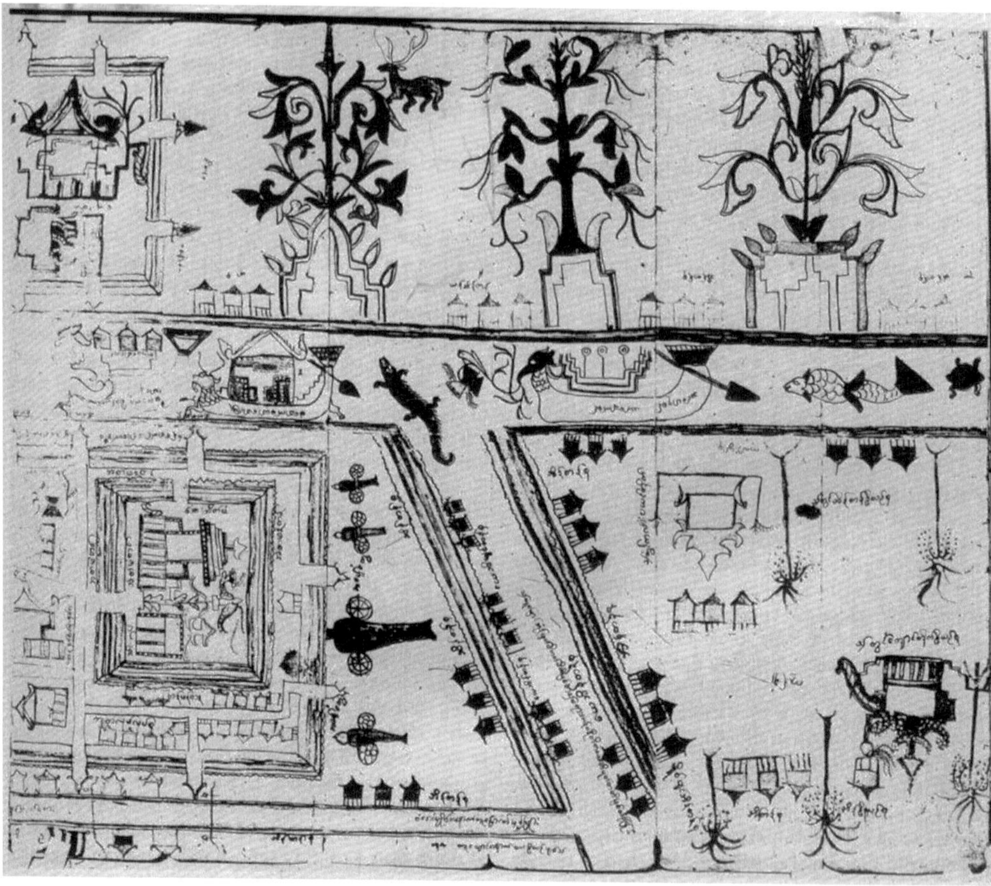

Figure 2: Detail of a large Burmese Military Map of Ayutthaya.
The pre-modern Thai capital, 1767.
By permission and courtesy of Thomas Suárez.

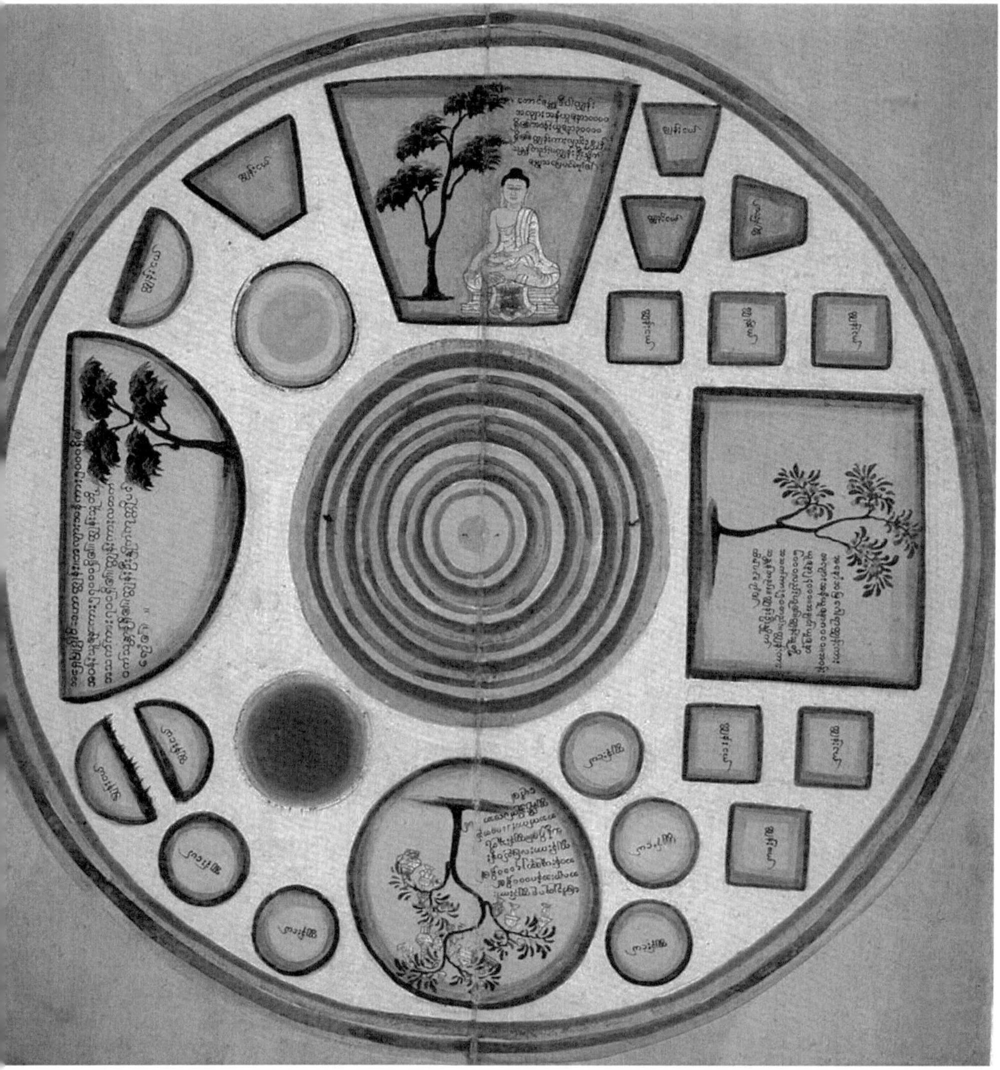

Figure 3: Burmese painting of the Cakkavāla.
A nineteenth century Burmese cosmological manuscript.
By permission of the British Library, London (or. 14004, fol. 27).

Figure 4: Shan map relating to a border dispute between (British) Burma and China along the Nam Mao River.
By permission of the Syndics of the Cambridge University Library.

Figure 5: Indian Jain Chart, *Manuslyaloka*, which is the generic title that refers to the "World of Man," fifteenth century.
By permission and courtesy of the Victoria and Albert Museum, London.

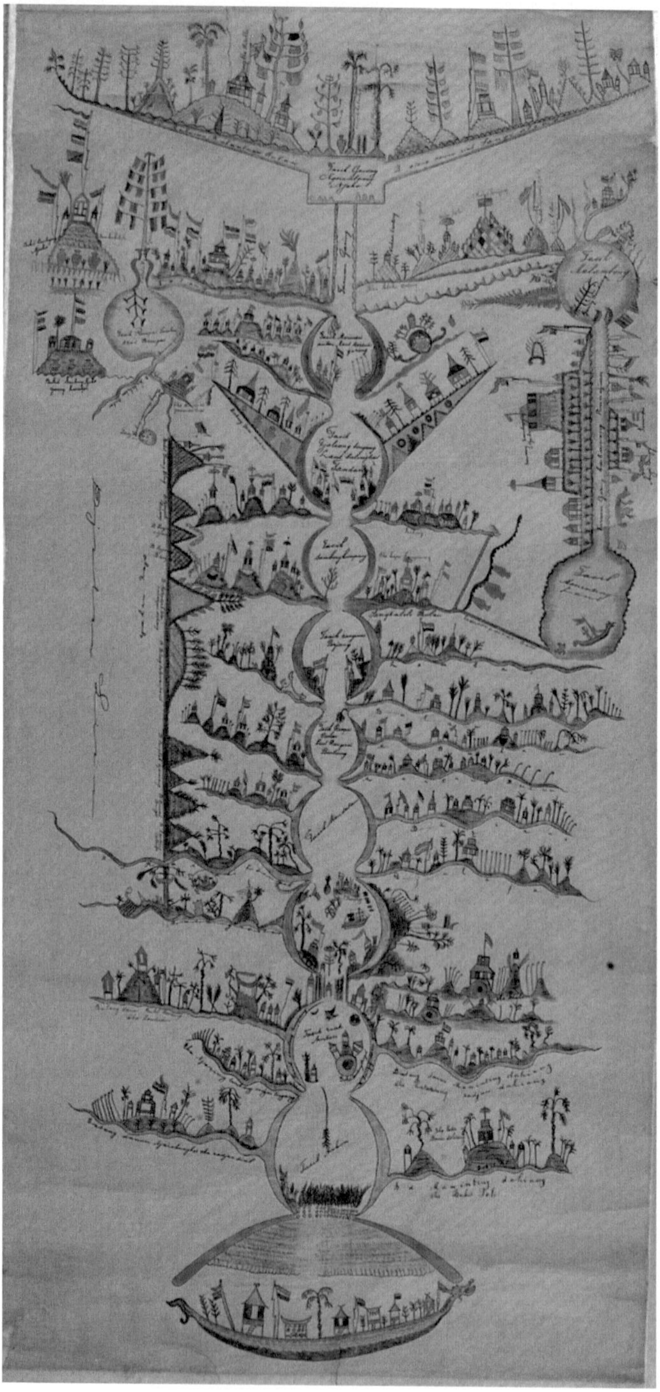

Figure 6: Ngaju Dayak Map of the Upperworld
From Southern Kalimantan (Borneo), early twentieth century map contains the notation *Djalan liau* (route of the souls of the dead).
By permission of Rheinisches Bildarchiv Koeln, Kattenbug, Germany.

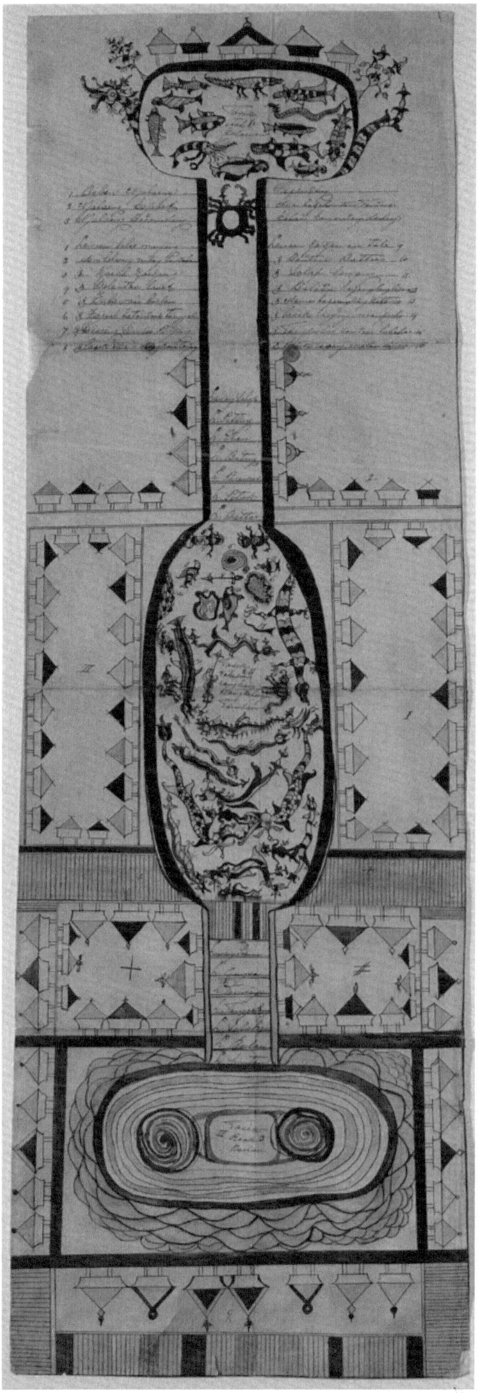

Figure 7: Ngaju Dayak Map of the Underworld
From Southern Kalimantan (Borneo), early twentieth century map contains the notation *Djalan liau* (route of the souls of the dead). By permission of *Rheinisches Bildarchiv Koeln*, Kattenbug, Germany.

Figure 8: Central Portion of the Sacred Map of Timbanganten.
Photograph courtesy and by permission of Joseph. E. Schwartzberg.

Figure 9: Antonio Saliba's 1582 Cosmological Chart
This map also references Tyco Brahe's comet in its eight circles. This cosmic model of concentric rings was derived from Ptolemy and Aristotle. This view of the earth represents a cross-section of sixteenth century understandings and world views from the perspectives of the physical, spiritual and conceptual.
By permission of the Herzog August Bibliothek, Wolfenbüttel: (K3.6)

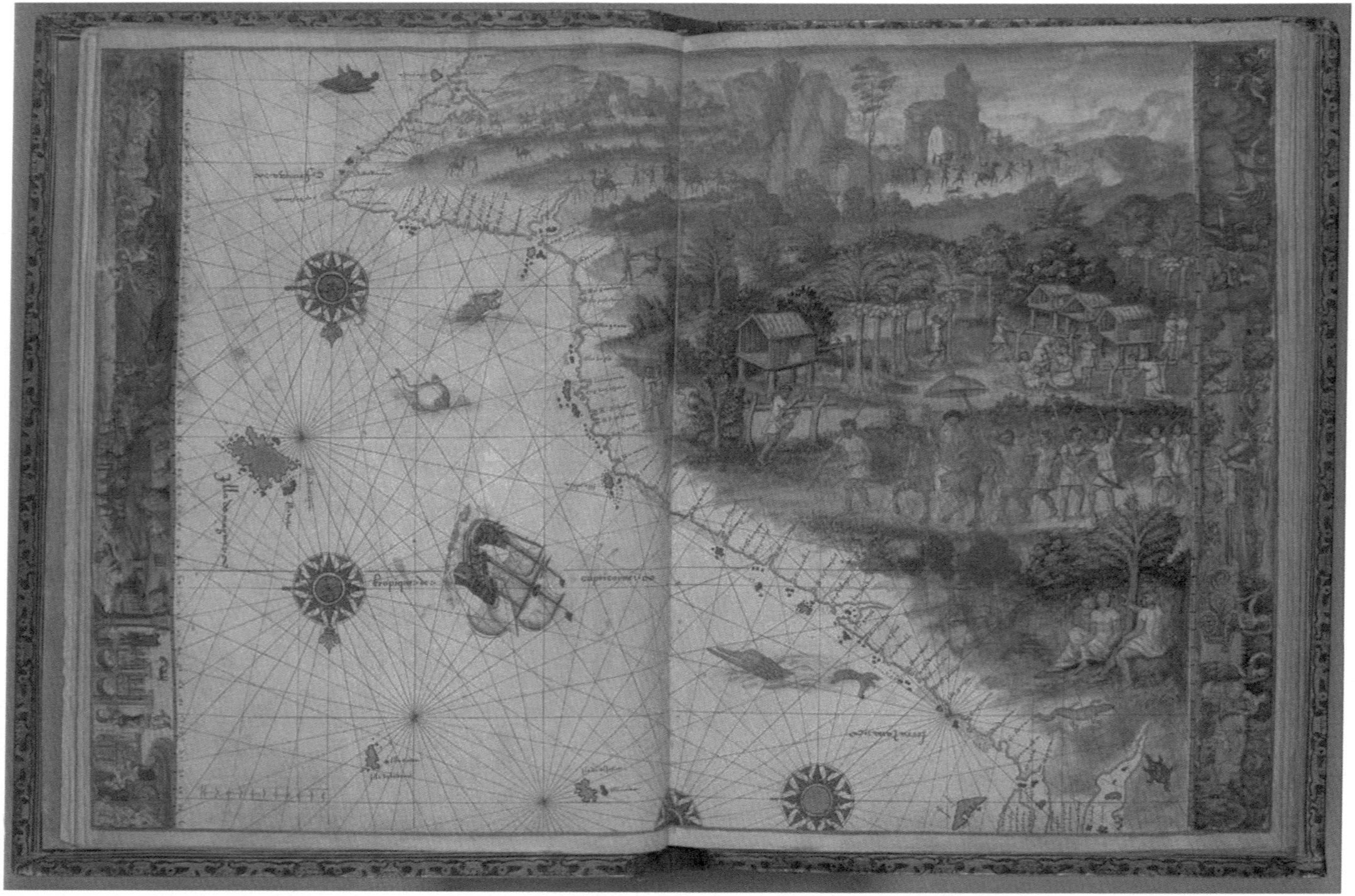

Figure 10: From the Vallard Atlas "Terra Java.." 1547; south is up. HM 29 (Map 1).
"This item is reproduced by permission of the Huntingdon Library, San Marino, California."

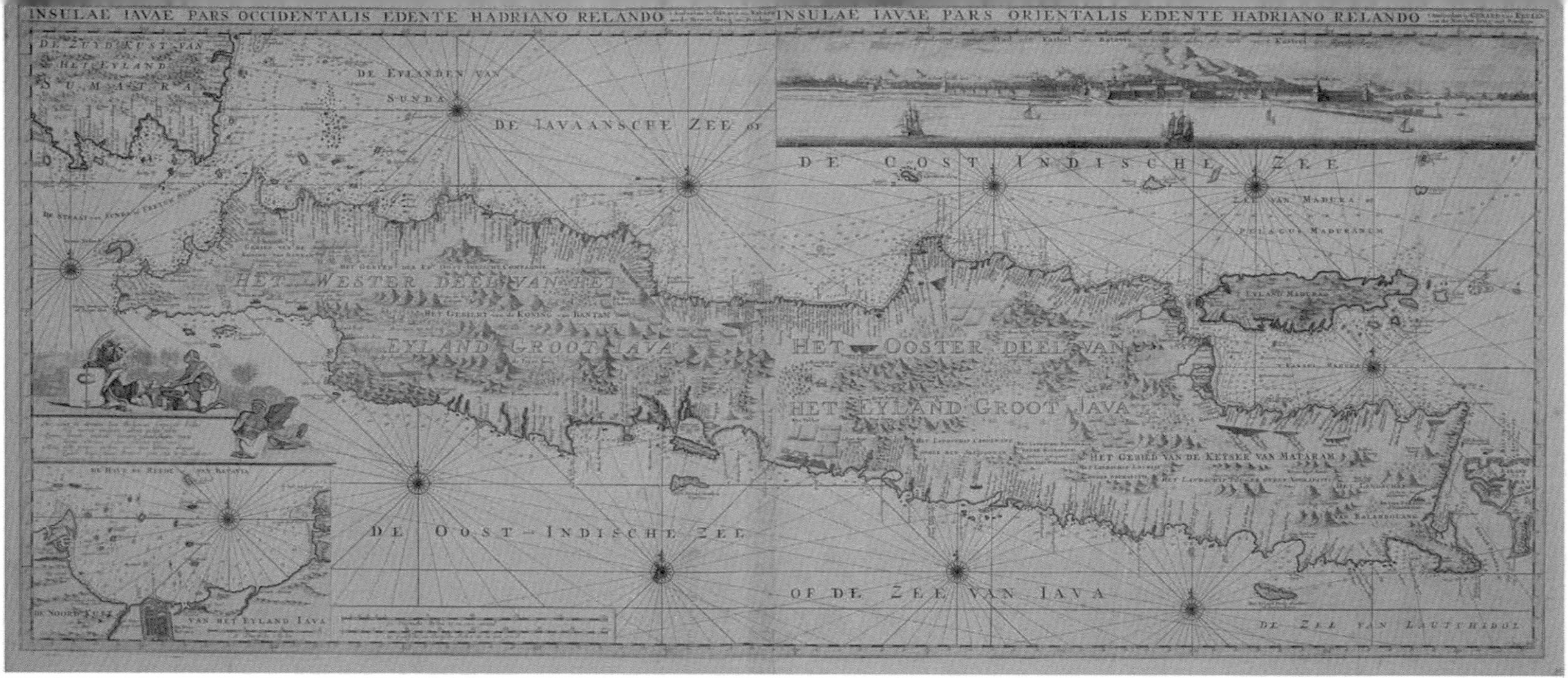

Figure 11:Map of Java by Gerard van Keulen 1728 (51 x 114 cm).
Insulae Javae Pars Occidentalis Edente Hadriano Relando. K193.
Collection of the Maritime Museum Rotterdam.
By the permission of the Maritiem Museum, Rotterdam.

Figure 12:Thailand and Burma Caricature Map. Frederick Neale, 1852.
This map was produced for some British officials on a visit to Thailand to explain a border dispute between the then Siam and Burma, As one can imagine they did not know quite to make of it.
By permission and courtesy Thomas Suárez.

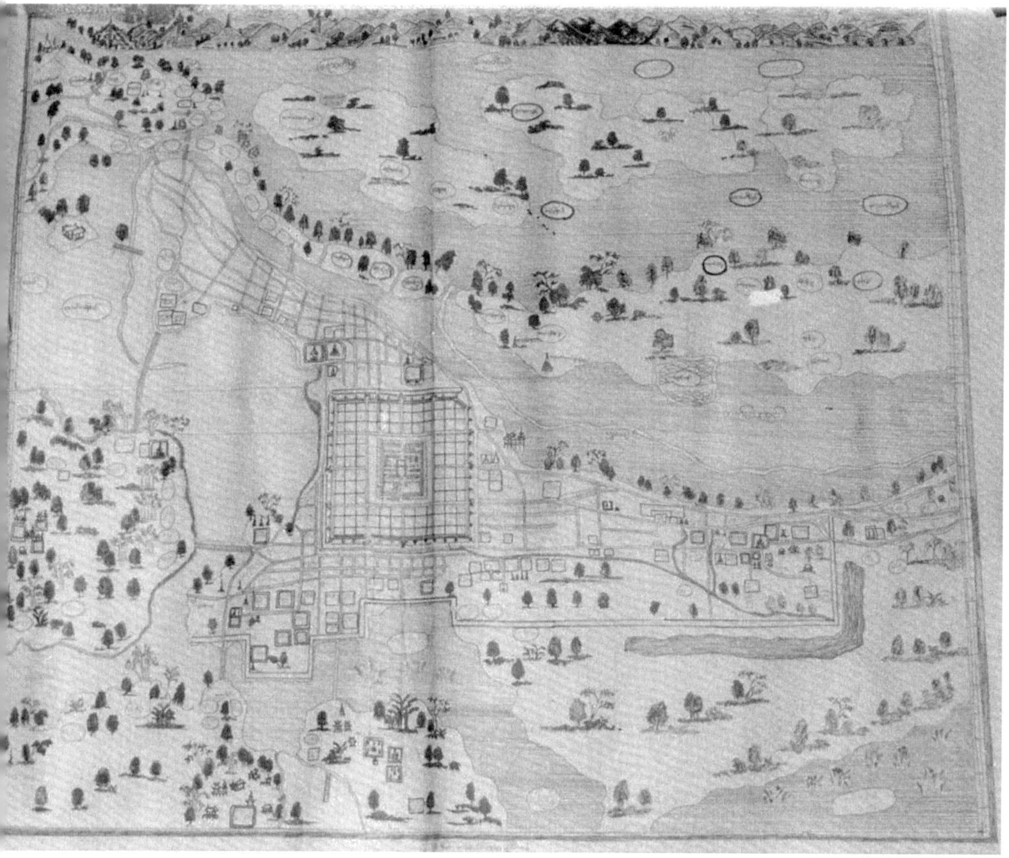

Figure 13: Map of Amarapura, Capital of Burma before the founding of Mandalay.

This is a copy produced by Burmese Monks at the Taung Lay None monastery in Amarapura, where it now is stored, of a mid-nineteenth original. 93.7 x 105.7 cm.

Photograph courtesy and by permission of Joseph E. Schwartzberg.

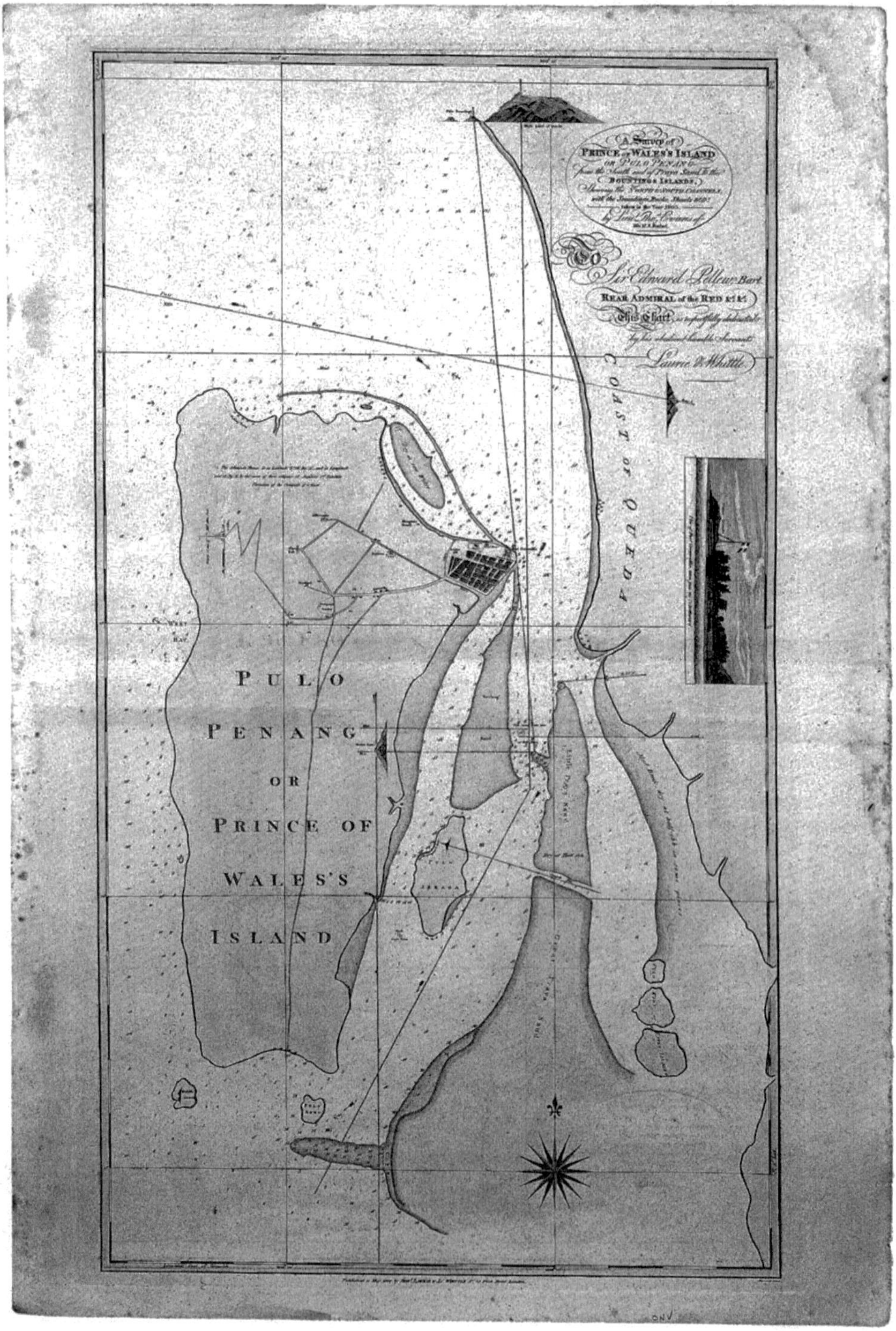

Figure 14: Prince of Wales Island or Pulau Pinang, 1807, Laurrie & Whittle. The British settlement of George Town is marked in red on the original colored map, on the eastern tip of the island.
By permission and courtesy of Thomas Suárez.

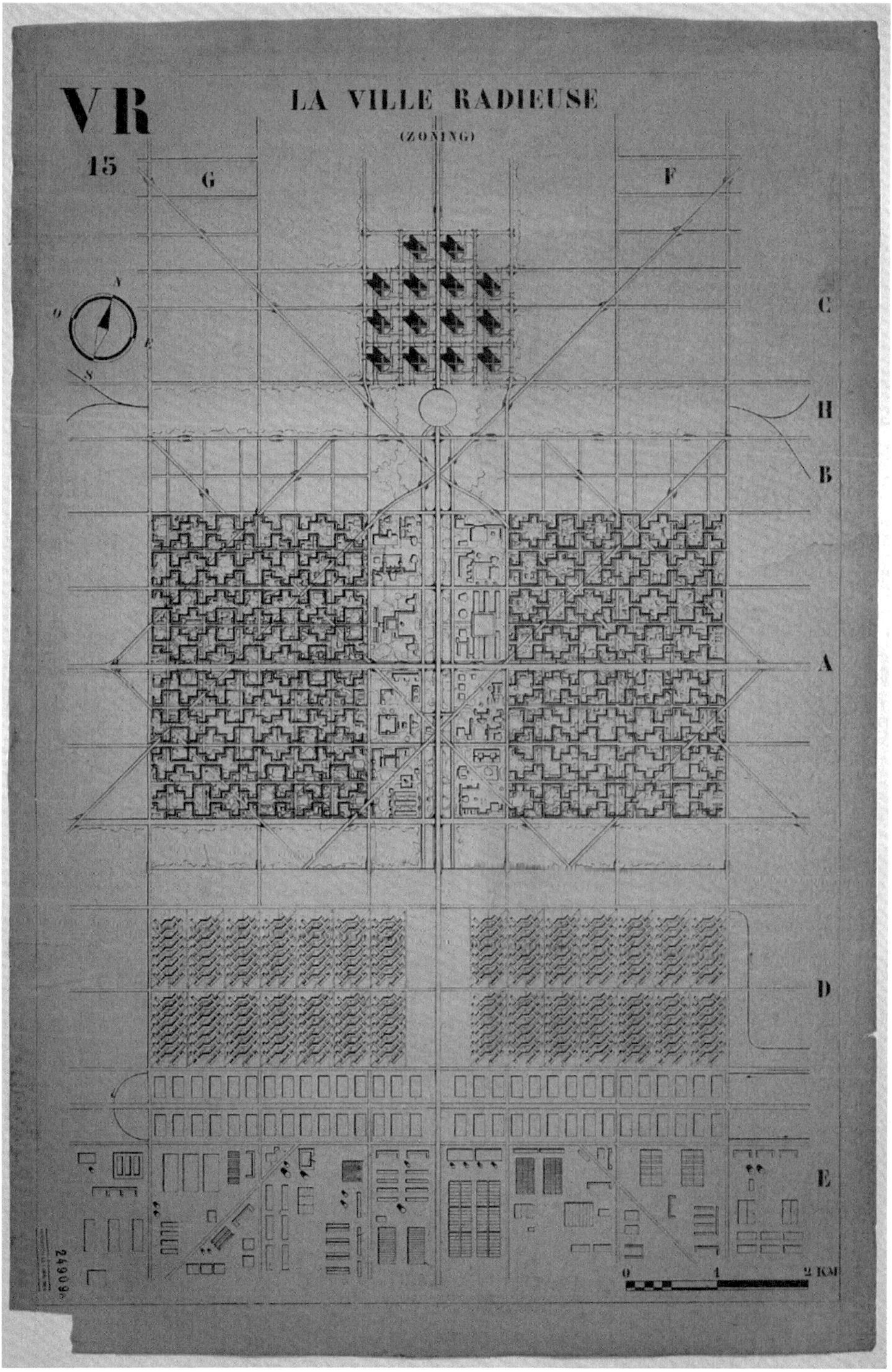

Figure 15:Plan of the Radiant City by Le Corbusier, 1930.
The central district is residential, above it the business district and below the industrial sector.
By permission of the Fundacion de Le Corbusier (Plan FLC 24909/ARS, 2013).

Figure 16:Food City: Agricultural, industrial and residential landscapes.
This image is to reflect the integrated relationships proposed between agriculture, industry and landscape in Richard Weller's concept for Perth 2050.
By permission and courtesy of Richard Weller.

Figure 17: Richard Weller's Garden City PODS, for Perth 2050, 2008.
By permission and courtesy of Richard Weller.

PART III: MAPPING AND MODERN URBANISM

CHAPTER SIX
Design and Rational Urbanism

As we have seen, particular modes of mapping reflected the different ways people interpreted their world at different times, and reflected particular qualities of human settlement. The map in the Enlightenment shifted from being a product of sociocultural, religious, and cosmographical concerns, and of a non-orthographic type, to a new rationalized, reductive, and planimetric format. The following discussion pinpoints another change in the nature of maps: a move to contemplating the world through theorized conceptions of space that also altered the map's influence in the production of urban environments. The forms of mapping that aided the translation of utopian ideas include the survey, master plan, and model. Utopian ways of thinking about settlement have been characterized by a change from a retrospective mode of engagement with one's surroundings to the projective.

In her acclaimed study *The Rule and the Model*, Françoise Choay traces an important shift in the conceptualization and production of human settlements. The shift was associated with the rise of modernity and marked by the emergence of written texts as a theorized mediation for the design and production of urban space. Until the dawn of modernity, she argued, there were no such texts:

> It is easy to forget that religion and the sacred have traditionally been the major factors organizing human space, either through the action of the spoken word or through the written word, which in the archaic world displayed commands of the gods on monuments

for all to see. We have forgotten that in societies without writing, the organisation of built space is interconnected with the whole of social practices and representations, without there having to be a single word in the lexicon dedicated to the reflection on the idea of spatial arrangement.[492]

Choay argues that the theorization of, and associated written discourse on, urban space is both recent and part of the domain of Western culture. It was instigated on a broad scale mainly because of the Industrial Revolution, upon which Western cultures imposed or persuaded the wholesale adoption of theorized urban concepts because of their credibility premised on a scientific basis.[493] The trend highlighted is the introduction of "a rational foundation" for urbanization and the texts, which assert the basis of rational urbanism. This trend assumes a scientific rationale while the authors of the theorized works convey a highly ideological position.[494]

The origins of the discourse on urbanism are traced to the treatise of Alberti, *De re aedificatoria*, published in the fifteenth century. Notably, Alberti does not refer to any religious reasoning for his method, although he highlights social activities of his era. Based upon mathematical ideas of the time and theories of perspective, "in proposing a rational method for conceptualizing and realizing buildings and cities," Choay writes, the treatise "successfully undertakes the task of establishing a relation to the built world unknown to Antiquity and the Middle Ages."[495] Alberti's work can thus be viewed as having laid the foundation for the discipline of urban theory, or what was referred to by nineteenth-century intellectuals as "urbanism."[496]

---

[492] Choay, *Rule and the Model*, 3.
[493] Ibid. See also Diane Brand, "Surveys and Sketches: 19th-century Approaches to Colonial Urban Design," *Journal of Urban Design* 9, no. 2 (2004): 153–75, for a discussion of the colonial mandate of an empirical approach for the cases of colonial Adelaide and Wellington in the nineteenth century.
[494] Choay, *Rule and the Model*, 2.
[495] Ibid., 4. See also Rocco Sinisgalli, *Leon Battista Alberti: On Painting: A New Translation and Critical Edition* (Cambridge: Cambridge University Press, 2011), a recent translation of Alberti's ideas predominantly in painting but also related to concepts for urban environments.
[496] Choay, *Rule and the Model*, 6. For Alberti's treatise, see Leon Battista Alberti, *On the Art of Building in Ten Books*, trans. Joseph Rykwert, Neil Leach and Robert Tavernor (Cambridge, Mass: MIT Press, 1988); Roland Freart de Chambray, *A parallel of the antient architecture with*

The groups of works contributing to conceptions of the built spaces are identified as "instaurational," and further divided into subcategories of "architectural treatises, utopias, and writings on urbanism."[497] These three divisions are key to understanding different ways of seeing, drawing, and mapping the world, and appreciating how rational urbanism came to the fore. Architectural treatises are defined by sets of principles and rules for arriving at built space. Choay dwells on the phenomena of literary utopias and their influence on built space; the focus here is the translation of utopian ideals into models and mappings, while analyzing their influence on urban design. A general definition of a utopia would be an imaginary, ideal place, exemplifying perfection in social, political, and cultural practices. Utopias exhibit "a critical approach to a present reality and the modeling in space of a future reality."[498] Thomas More's utopian literary version is the precedent, whereby the arrangement of space is achieved through modeling, and this methodological process unleashes numerous potentials for imagining and ultimately altering the natural world.[499] More's character Raphael, as the witness, first describes utopia to the readers to convince them, and then suggests that utopia is not only a place but also a prototype. However, the model or prototype aspect of utopia is that it appears "delocalized and reproducible."[500] It is a phenomenon solely focused on the rules of an established human order as well as a rigid system of sociocultural practices.[501] In terms of representation, these rules may be relayed via master planning, diagramming, or as a three-dimensional prototype. Finally, the group of texts known as "writings on urbanism" demonstrates how writings about concepts of rules and models have been transformed over time. Despite their limitations, reinterpretation and resistance to "the restructuring of knowledge, and bearing a meaning that transcends their content," they survive to influence conceptions of built space even today.[502]

---

the modern, in a collection of ten principal authors who have written upon the five orders, 2nd ed. (London: Dan Brown, John Walthoe, Benjamin Tooke & Daniel Midwinter, 1707).

[497] Choay, *Rule and the Model*, 8; Thomas More, *Utopia* (Adelaide, South Australia: The University of Adelaide Library, 2008) http://ebooks.adelaide.edu.au/m/more/thomas/m83u/.

[498] Choay, *Rule and the Model*, 7–8.

[499] Ibid., 137.

[500] Ibid., 139. See also Colin Fournier, "Webbed Babylon," *Architectural Design* 71, no. 3 (2001).

[501] Choay, *Rule and the Model*, 139.

[502] Ibid., 9.

The identification of a series of divisions in the theorization of the built environment is, in effect, an exposition of the forms of rational urbanism associated with the advent of modernity. The analysis of urban theorizations extends to the concept of mapping and mapped representations of built space. Replacing non-theorized practices of spatial construction with highly conceptualized methods of building is the beginning of urban rationalism, which also signifies a distinct shift in mapping and conceiving urban space.

6.1 Retrospective and Projective Mappings

Retrospective and projective mappings refer to two related, yet distinct, ways of dealing with the mapped environment. One is descriptive and concerned with the past; the other is speculative and concerned with the future. All forms of premodern mapping discussed in earlier chapters were retrospective in nature; they were produced to describe or narrate a story about an event, site, or idea. They were not maps for design purposes. Design maps are projective. They describe a future reality through conceiving, planning, and projecting for the purpose of implementation in the field. Retrospective and projective mapping, considered in the context of utopian thinking, may be seen as resonating with the "culturalist" and "progressivist" utopian approaches, as proposed by Choay. The former suggests reverting to past modes of production, whereas the latter looks to the machine age for insight and motivations.

An historical study of the agency of the map, or what may be considered a retrospective mapping, looks back at people's way of life—their sociocultural practices, religious influences, and technologies—to come to an understanding of how they described or constructed their worlds through maps. Creation of built environments happened spontaneously because of need rather than because of a preconceived holistic planning exercise. In this context, premodern maps could not be considered design tools, in the modern sense, nor integral to the design process. By contrast, when we use mapping as a projective device, the map becomes an instrument of design and is utilized with an understanding that there are a series of rules pertaining to the conception of settlements. For example, in the process of surveying, maps are constructed by codes, instruments, and techniques that only create a representation of reality possible through those techniques.[503] Conditions put forward by this initial contact with the site are prioritized and others ignored to frame the design space.[504]

Typically, the design process involves looking retrospectively to be better informed about the context within which the urban design will be realized. The point at which this process becomes contentious is where the shift occurs from the retrospective to the projective position. How does the designer employ traditional and historical knowledge of an environment to undertake a projective mapping of the new urban reality? The junction between the traditional and the contemporary is often uncomfortable and unexplained because, in this transition, the designer usually views and interprets historical information from a perspective unrelated to any of the contextual or intellectual scenarios that gave that historical material meaning. Thus, there is an inherent disjunction between the retrospective and projective reading as well as mapping in the process of design. This disjunction in the design process has in many ways contributed to unsatisfactory design outcomes, which include the kitsch and pastiche.[505] The focus of this study is to gain an understanding of the function of the map in urban design.

Central to the emergence of rational urban planning was utopian thinking and its translation into a visual representation being centered on a map or a plan. If we view urban rationalism from the utopian perspective, while focusing on the map and mapping practices, three stages can be identified in the process of its development. The first is the transitional stage, when intellectuals from two differing perspectives— retrospective and progressive, embraced utopian thinking. Nineteenth-century utopian literary thinkers are grouped into "culturalist" (William Morris), which aligns with the concept of looking retrospectively, while "progressivist" approaches (Robert Owen, Charles Fourier) involve looking to the future for inspiration rather than the past. As far as authors and architects composing text for the explicit purpose of influencing urban and mapped space are concerned, Camillo Sitte was considered part of the "cultural organicity" of the utopian movement, whereas Le Corbusier, as a progressivist, focused on "progress and rationality."

---

[503] Corner, "Agency of Mapping," 216. See also Brand, D, 2004.
[504] Corner, "Agency of Mapping," 216.
[505] See texts such as Daniel Solomon, *Global City Blues* (Washington, DC: Island Press, 2003).

The second stage in the development of urban rationalism involves the reduction of utopian progressivist concepts into design principles and rules. These parameters of design may also be considered new forms of mapping. Two kinds of maps most affected by the rules stemming from utopian thinking are the survey and the master plan. The survey, as a representation of geographical data, relies on sets of decisions, observations, and interpretations to arrive at the data presented—a highly rationalized process. The greater the perceived accuracy, the greater its scientific merit leads to the distinction of the survey as a true representation of geographic reality and a credible tool for analyzing the site. The survey also became part of a greater mapping project, which symbolized the advent of modernity in the quest for imperial expansion and the rise of colonial cities, as was discussed in Part II. The survey, according to James Corner, is considered "quantitative, objective and rational, it is also true and natural." Therefore, it actually aids in the production and marketing of the future design plans for the settlement. The master plan of human settlement is the other form of mapping, which is a graphic representation of utopian objectives. This mapping specifies, through block sizes, orientation, setbacks from site boundaries, access routes, configurations' zoning requirements, and built envelope parameters, the type of development required to satisfy the new mandate for a new way of life. It is an assumption that the map's role is to "objectively identify and make visible the terms" by which the new design proceeds.

The third stage in the development of the rationalism process highlights a metaphorical leap between conceived design principles and the physical design implementation. This is the realization of the "model," whereby "it offers, on the level of the imaginary, a device for the a priori conception of built space." The model represents yet a further reduction of progressivist utopian concepts because of its format as a physical representation. A model in the urban environment usually has three-dimensional parameters and, therefore, in every dimension, there is less scope for interpretation and adaptation by those applying it in the field. The final stage of urban rationalism is the actual implementation of the model and rules of construction into a physical reality.

Utopia: Projective and Retrospective

Jean Baudrillard suggests that to distinguish between what is *real* and what is *represented* is no longer meaningful.[506] Corner believes that to distinguish completely between the "real world" and the constructed one would not only be unimaginative but act contrary to human nature's desire to "structure reciprocal relationships with its surroundings."[507]

Utopianism is a key aspect of critical thought related to urban environments and living conditions. David Pinder argues that dissatisfaction and desire for change means utopian thinking is instrumental in "developing critical approaches to cities and processes of urbanization."[508] As has been foregrounded by Choay, utopian thinking stimulated a new way of approaching ideas about urban environments. Premodern modes of production of built space involved established patterns of life based on socioreligious influences and practices, local technologies, and sets of culturally specific rules, with place-making occurring organically because of quotidian existence. However, as modern urban practices emerged, settlements became preconceived, and the advent of theorizations and premeditated plans were used for achieving a better way of living. The mapping of these new imaginings of space, via their reduction into rules of practice, signals the beginning of the rationalist turn. In early utopian writings, the rationalist turn is not evident. It is only when the thinking needs to be converted to something tangible—able to be drawn or universally understood for implementation—that the utopian elements to the theorizations appear compromised.

Choay's definition of the utopia of literary projections describes the function of model society as follows:

> The model society is opposed to a historically real society, and the criticism of the latter is indissociably linked to the description of the former; the model society is supported by *a model space, which is an integral, necessary part of it*; the model society is located

---

[506] Jean Baudrillard, *Simulcra and Simulation*, trans. Sheila Faria Glaser (Ann Arbor: University of Michigan Press, 1994), 2. See also Jean Baudrillard, *Utopia Deferred: Writings for* Utopie *(1967–1978)*, trans. Stuart Kendall (New York: Semiotext(e), 2006).

[507] Corner, "Agency of Mapping," 222.

[508] Christian Hermansen Cordua, "Prologue," in *Manifestoes and Transformations in the Early Modernist City*, ed. Christian Hermansen Cordua, 1–5 (Farnham, Surrey: Ashgate, 2010), 2.

> outside of our system of spatio-temporal coordinates, it is *elsewhere*; the model society is not subject to constraints of time and change.[509]

Thomas More's *Utopia*, as the first writing to exemplify the above-mentioned characteristics, is the foundational text in Choay's discussion. However, there were also different kinds of views purveyed within the genre of utopian writings. The group identified as culturalist or retrospective identifies with different values systems compared with those of the projective or progressivist group. The former includes writers such as William Morris, A. W. N. Pugin, and John Ruskin, who aspired to past modes of production related to the craft movement, and advocated the rejection of the machine and industrialization.[510] Camillo Sitte also promoted the quality of spaces of past traditional European cities. However, Sitte was realistic about the role of industrialization and its effect on the city because "his aim was not a recreation of historic city patterns but the improvement of the grid and block system with minor displacements and additions."[511] Sitte's master plan frames squares with borders to create smaller squares, which are aspects of traditional cities. However, a key aspect of the Privoz master plan by Sitte is its attention to detail and craft-like quality that, it may be argued, reflected the value systems he applied to urban space. In essence, the culturalist approach typical of Sitte was also seen in Viollet-le-Duc's work. They shared similarities in their writings and conceptions of the new utopian city, from a social approach, its rationalism, and pedagogy, as well as their criticism of the drawing board dominating the mode of production of urban environments. The culturalist approach preferences the use of traditional urban morphologies.[512]

Sitte was inspired by picturesque drawings, and baroque planning traditions influenced his urban design solutions. The master plan for Privoz highlights an arrangement of axial vistas to the cathedral, culminating with a

---

509 Choay, *Rule and the Model*, 34.

510 David Pinder, "Utopian Urbanism: Ideals, Practices and Prospects," in Cordua, *Manifestoes and Transformations*, 10.

511 Ruth Hanisch, "Camillo Sitte: City Planning According to Artistic Principles, Vienna 1889," in Cordua, *Manifestoes and Transformations*, 128.

512 Hanisch, "Camillo Sitte," xiii.

public space in front of the monumental building. Sitte's urban design and representations may be seen as transitioning between classical space and modern space.

By contrast, Le Corbusier represents the projective approach to utopia, prioritizing the machine and industrialization in determining a new philosophy for the production of spaces for living. These values were evident in Le Corbusier's representations in their single clean lines, perspectival and axonometric projections as well as the monumental scale of his urban schemes. Projective utopian visions were typically based upon belief systems, which have faith in "the power of space to shape human activities and their projections of an ordered space become a means to contain social process and change, and to ensure harmony and stability."[513] Frank Lloyd Wright, Le Corbusier, and Ebenezer Howard were among the earliest of the modernist utopian planning movement.[514] The rational style of modernism was first highlighted through philosophies put forward by the New Objectivity movement, of European origin from the 1920s, as well as in their choices of projection for their large-scale urban visions.

Designs featured highly detailed yet simple forms with clean lines and devoid of any decorative or ornamental treatment. Steel and concrete were the materials of choice for architects such as Ludwig Mies Van Der Rohe and Walter Gropius because these austere materials reflected the "progress and prosperity" of the machine age.[515] These characteristics of modernism were transferred to realizing urban environments of gridded and regularly arranged street systems. This level of organization of the infrastructure, parklands, and vehicular and pedestrian traffic for the city facilitated efficiency of movement, capital, and transport systems, maximizing production and output.

Howard Frederic Osbourne, Clarence Stein, Lloyd Wright, and others believed such regularization of building and urban forms would bring about the

---

513 Pinder, "Utopian Urbanism," 20. See also David Harvey, *Spaces of Hope* (Edinburgh: Edinburgh University Press, 2000), 160; F. Luckerman and P. Porter, "The Geography of Utopia," in Lowenthal and Bowden, *Geographies of the Mind.*
514 Robert Fishman, *Urban Utopias in the Twentieth Century: Ebenezer Howard, Frank Lloyd Wright, and Le Corbusier* (New York: Basic Books, 1977). See also Leonie Sandercock, *Towards Cosmopolis: Plannning for Multicultural Cities* (Chichester: Wiley, 1998); Leonie Sandercock, *Cosmopolis II: Mongrel Cities of the 21st Century* (London: Continuum, 1998).
515 Pinder, "Utopian Urbanism," 166.

demise of social malaise, which had come to dominate urban environments from the inception of large-scale industrialization.[516] Ebenezer Howard outlined planned steps to achieve his vision for the city, under the title of the garden city movement. His model intially set out to map social reform for cities but eventuated in realizing environmental reforms.[517] Le Corbusier and CIAM, established in 1928, went on to adopt the Athen's Charter as their sets of design rules; they identified urban planning as[518]

> "the cause of the anarchy that reigns in the organization of cities and the equipment of industry", and in response demanded large-scale urban reconstruction and the insertion of new spaces based on strict zoning between functions.[519]

Thus, utopian thought digressed from realizing social reforms to focussing on issues of planning and regulating through sets of rules. The richness of historical readings as well as the complexity of utopian visions was reduced to patterns and rules as sets of design tools. To give the design outcome legitimacy, the process relies on conviction derived from theorizations. However, the projective process remained in actuality more or less a series of decisions around individual preferences. Le Corbusier's preferences for modernist living involved unprecedented density and scale, and the effects of these design choices were readily conveyed in his planning and perspectival drawings.

---

[516] John A. Hannigan, 1995. "Postmodern Architecture and Urban Design," *Current Sociology* 43, no. 1 (1995): 167.

[517] Pinder, "Utopian Urbanism," 19. See also Ebenezer Howard, *Garden Cities of To-morrow (Being the Second Edition of "To-Morrow: A Peaceful Path to Real Reform")* (London: Swan Sonnenschein, 1902).

[518] CIAM is the acronym for *Congrès Internationaux d'Architecture Moderne* (International Congresses of Modern Architecture) founded in 1928 and disbanded in 1959. Its main objective was to spread the principles of the modern movement focusing on the main domains of architecture, landscape architecture, urbanism, industrial design and others. http://encyclopedia.thefreedictionary.com/Congr%...

[519] Pinder, "Utopian Urbanism," 19. See also Le Corbusier, *The Athens Charter* (New York: Grossman, 1973).

6.2 Mapping and Designing

Corner argues that contemporary planning practices are preoccupied with tools from Enlightenment and modernist paradigms.[520] Contemporary projects often engage with the site as *tabulae rasae* or as geometrically manipulated from above, rather intrinsic to the design process.[521] It is evident how the two predominant forms of mapping became instrumental in translating projective utopian concepts into formal rules and principles of design: the *survey* and the *master plan*. The ways in which the survey and the master plan were created and used reveal the influence of Enlightenment rationalism. Surveys were considered quantitative inventories, and master plans reflected rational and contained ideals. A good example of this practice is evident in Le Corbusier's *Voisin* plan, in the combination of the aerial photo and the shaded plan. The geometrical design in the master plan solution set against the context of the old town is overwhelming, given that they are projected at the same scale.

Le Corbusier's belief, for example, envisaged centralized planning as the solution not only to urban buildings but as infiltrating every aspect of life, in particular, as an alternative to France's urban ills.[522] The *Voisin* plan in its juxtaposition with the aerial photograph also suggests that an urban malaise of jumbled old urbanities may be removed with the ordered new forms of planning. The drawing is key to marketing this message, as Le Corbusier further articulates the power of planning as follows:

> In this system, everything would be determined by the plan, and the plan would be produced "objectively" by experts; the people who would have a say only in who was to administer it. "The harmonious city must be first planned by experts who understand the science of urbanism. They work at their plans in total freedom

[520] Corner, "Agency of Mapping," 224. See also Eugene Halton, "The Modern Error: Or, the Unbearable Enlightenment of Being," in *Global Modernities*, ed. Mike Featherstone, Scott Lash and Roland Robertson (London: Sage, 1995).

[521] David Pinder, *Visions of the city: Utopianism, power and politics in 20th century urbanism.* (Edinburgh: Edinburgh University Press, 2005) 72. See also Francesco Careri, *Walkscapes: Walking as an Aesthetic Practice* (Barcelona: Aleu SA, 2003).

[522] Peter Geoffrey Hall, *Cities of Tomorrow: An Intellectual History of Urban Planning and Design in the Twentieth Century*, rev. ed. (1988; rev. Oxford: Blackwell, 1996), 210.

from partisan pressures and special interests; once their plans are formulated they must be implemented without opposition."[523]

The so-called science of urbanism, proposed by Le Corbusier and other modernists, in its projective mode of design, presented an imposition of utopian ideas onto urban environments that seemingly underestimated the sociocultural complexities of urban life. The graphic translation of modernist ideas, such as the example of the Contemporary City, illustrates the monumentality and highly rationalized nature of these schemes. Although Corbusier intended the representations to symbolize order and harmony, perhaps they were also seen by governing bodies as annihilating every trace of the past and any established sociocultural practices or qualities of context, because in many instances they were not built.

CIAM's philosophy summarized town planning into four functions consisting of living, working, transport, and recreation.[524] These themes aided the translation of original utopian agenda into diagrams for persuading, coercing, and dictating to "experts" (presumably, the planners and architects), and representing the form of urban settlement, which was required to fulfill utopian objectives in order to achieve a new way of living. They also contributed to the realization of the utopian model.

Diagramming of projective utopian concepts, which was understood by 1960s urbanists as legible and easily negotiated urban space, resulted in the evolution and future resilience of the modern master plan of human settlement. The reduction of concepts, which were intended to reshape human behaviors, and enhance sociability and the efficiencies of the functioning of everyday urban life, to diagrammatic models of spatial interactions prompted criticisms that the city had lost its meaning.[525] Lefebvre concurs:

---

[523] Le Corbusier, *Athens Charter*, 239.

[524] Mark Wigley, *Constant's New Babylon: The Hyper-Architecture of Desire* (Rotterdam: 010, 1998), 131.

[525] John Urry, *Consuming Places* (London: Routledge, 1995). See also Libero Andreotti and Xavier Costa, *Theory of the Derive and Other Situationist Writings on the City* (Barcelona: Museu d'Art Contemporani de Barcelona / ACTAR, 1996).

> The structures of space cannot be explained by a vitality of an urban community. They are the result of history that has been understood as a creation of social agencies of actors, generally of collective subjects. From their interactions, their strategies, successes and failures, derive the qualities and characteristics of urban space.[526]

The emergent modern mappings graphically demonstrated value systems to reshape society. For example, modernist public space in its scale and exposure was more easily monitored compared with spaces of traditional cities. As representations of social reform, modernist mappings in their projective utopian nature prioritized the ordering space for ordering society. Therefore, they may be seen as an apparatus of government as well as tools for design.

6.3 Mapping as Designing

When we use mapping as a projective device, conveying a modeled and highly defined spatiality, the map becomes a design tool and is utilized with an understanding that there are a series of rules pertaining to the conception of settlements. Other graphically inspired "tools of thought" included modes of projection, such as axonometry, orthography, perspective, quantitative surveys and inventories, which were all modes of representation used in the dissemination of modernist works, such as Le Corbusier's single line perspective of the Radiant City.

The modernist city became the radical new model or, arguably, a three-dimensional mapping of the post-industrial age. It served as both a conceptual and a physical model for urban form, and through the implementation of regular geometric forms and minimal aesthetic embellishment of spaces, it was intended to alter human behavior and ways of living for the common good.[527] In their geometric regularity and imposed sense of order, modernist maps of urban space were expected to have social ramifications, such as social harmony, order, and

---

[526] Henri Lefebvre, *Die Revolution der Städte* (Frankfurt: Syndikat, 1976), 137. See also Simon Sadler, *The Situationist City* (Cambridge, MA: MIT Press, 1998).
[527] Pinder, "Utopian Urbanism," 20.

control.[528] The master plan governs from above, as a source of graphic instructions indicating how the settlement should function and organize its peoples.

The modernist city, influenced by the nature of industrialized cities' urban slums, invoked radical change of urban forms in opting for high-rise buildings for residential accommodation in order to prioritize urban park land space. High-density residential towers would be able to cater for large population increases, which were occurring as people moved from agrarian-based activities to the vocations offered by an industrializing city.[529] Catering for increasing urban populations was also a key component of Le Corbusier's motivations for realizing the high-rise city. The choices of different typologies for living, structured, and geometrical forms equated to new urban design thinking.

Le Corbusier famously coined his paradoxical position in the realization of his new visions for cities and said, "we must decongest the centers of our cities by increasing their density."[530] Le Corbusier's new principles of the city were intended for a cleared site, whereby city centers would be demolished for large boulevards and high-rise towers. This functional reorganization was also to correspond with a reconfiguring of accepted social structures. Le Corbusier stipulated that the "cell" apartments would be all mass-produced in keeping with the period of industrialization. Variations would only occur in the layout of blocks and planting types in the green spaces at ground level.

The austerity of such proposals was reflected in the nature of the design tools—the survey and master plan—and evident in the modeled space. The value systems associated with designers of cities of industrialization shared the view that city form had to respond to the efficiencies of production characteristic of the

---

528 Christian Hermansen Cordua, Manifestoes and Transformations in in the Early modernist City. (Surrey, England: Ashgate, 2010), 20.

529 Typically, a modernist vision for the city entailed "space shaped according to an overarching social objective; architecture and planning aim to integrate the metropolis; imposes an external utopian vision; commerce-oriented; geographically centralised; austere, inflexible; authoritarian; rectangular, unadorned." Hannigan, "Postmodern Architecture," 165.

530 Hall, *Cities of Tomorrow*, 207. For the political view of the role of utopian cities, see Frederic Rouvillois, 2000. "Utopia and Totalitarianism," in *Utopia: The Search for the Ideal Society in the Western World*, ed. Roland Scaher, Gregory Claeys, and Lyman Tower Sargeant, 316–32 (Oxford: Oxford University Press, 2000), 316; Russell Jacoby, *The End of Utopia: Politics and Culture in an Age of Apathy* (New York: Basic Books, 1999), xi–xii.

age because industrialization was a permanent condition. The modes of representing the functional requirements of human settlement at this time were also reduced down to their most rational forms, metaphorically championing industrial processes, but also responding to the bleak social and living conditions industrialization had created in traditional urban environments (refer Figure 15). Further improving on the legacies of Enlightenment technologies for perceived accuracy and precision, the survey and master plan facilitated a new mode of urban design, and the model form provided a three-dimensional translation of these ideals, which led to the creation of modern cities.

6.4 Maps, Society, and Urban Life

The ultimate dissemination of the model is through projective utopian approaches, which are the most common form of appropriating urban space for the colonial city. Architects such as Le Corbusier and Cerda saw the opportunities for realizing their new ways of living, and social agendas for change, in the colonial settlements of India and South America. The recipients of colonization and their local traditions, customs, and specific ways of living were largely disregarded for the priority of installing experimental model utopian urbanities. The projective utopian approach was legitimized as "an integral part of regional masterplans even though it denatures and dehumanizes the surface of the earth through the abstract projection of built forms."[531] However, urban rationalism, according to Choay, has been sustained in colonial policy ever since the sixteenth century, in that "the pattern for colonization of the Third World, in which industrialization reigns by means of the modeling of its spaces."[532] Utopian models typically adopted on a broad scale suffered the fate of the utopia presented by Thomas More. More predicted control and conditioning of public behaviors, in the creation of homogenous and stereotypical urbanities would be detrimental to the citizens of utopia. The argument for implementation of rational urbanist projects into colonized regions was typically one of efficiency, in terms of economies of scale and time by the standardization of urban space, which in reality proved quite the opposite.

---

[531] Hall, *Cities of Tomorrow*, 277.
[532] Ibid.

Le Corbusier's opportunity for realizing his only built urban plan, at Chandigarh in India, demonstrated the problems of translation of utopian thought. Lofty ideas converted to geometrical rules and principles to realize the model resulted in a severely compromised product. The survey and master plan were also integral to the type of urban form produced. Sandercock recounts, "modernist thinking about the city tended to see rational order in visual and aesthetic terms…" prioritizing orthogonal space.[533]

Chandigarh demonstrated for Le Corbusier a shift from a focus on implementation of a certain planning style and formula synonymous with his CIAM-based agenda to the prioritization of an architectural style. As Sarin suggests, Le Corbusier demonstrates an ideological shift

> towards a preoccupation with visual form, symbolism, imagery and aesthetics rather than basic problems of the Indian population. By concentrating on providing Indian architecture with forms suited to the Second Machine Age, the existing Indian situation could be more or less ignored.[534]

Despite proposed overarching social and environmental objectives, projective modes of utopian design once translated from rationalized sets of rules and principles into a physical form failed to deliver reform for the case of Chandigarh. Manfredo Tafuri suggests that the technical revolution, which occurred through industrialization, permitted new "laws of production." This process in turn enabled "new conventions" to be embraced.[535] Tafuri states that "architecture as ideology of the plan is swept away by the *reality of the plan* when, the level of utopia having been superseded, the plan becomes the operative mechanism."[536]

The client for Chandigarh was the postcolonial government of the autocratic British Raj, and the design consisted of a gridded sequence of roads,

---

[533] Sandercock, *Cosmopolis II*, 29.
[534] M. Sarin, *Urban Planning in the Third World: The Chandigarh Experience* (London: Mansell, 1982), 47.
[535] Ibid., 35. See also Manfredo Tafuri, *Architecture and Utopia* (Cambridge MA: MIT Press, 1976).
[536] Tafuri, *Architecture and Utopia*, 135.

building forms and layouts, which adhered to European climatic principles rather than north Indian. Disregard for sociocultural practices particular to an Indian way of life were evident in Chandigarh, and as Evenson argued, building layouts demonstrated a "total failure to produce built forms that could aid social organization or social integration; the sections fail to function as neighborhoods."[537] The omission of qualities integral to Indian society highlights the extent to which the power of rationalized planning rules and models were exonerated in favor of basic human spatial qualities of settlement.

As Leonie Sandercock states, this belief in scientific method stemming from Enlightenment understandings, which privileges the technical over any other forms of knowledge, dismisses a realm of equally important knowledge systems: "experiential, intuitive, local knowledges based on practices of talking, listening, seeing ... visual and other symbolic [representations]."[538] According to Eunice Seng, not all European modernist interventions into colonial space had such negative effects. When the housing estate of Tiong Bahru was proposed in Singapore in 1948, there were clear adaptations of European modernist concepts to suit the local sociocultural requirements. As Seng explains, Tiong Bahru exemplifies garden city utopian planning, where the green space was intended to be the rehabilitation area for the mind and body.[539] Seng highlights that there was a flexible arrangement put in place to cater for the local citizens. A green belt of open space around Tiong Bahru was reserved with the intent of it being used for recreation or agricultural purposes, and building development such as urban kampongs was prohibited. Over time, these unprogrammed spaces became zoned as spaces of communal social amenities.[540]

As Corner suggests contemporary designers, view space as being more complex than just physical objects and forms.[541] Corner believes instead that "inter-relationships" and the effects of "dynamic interactions" characterize

---

[537] N. Evenson, *Chandigarh* (Berkeley and Los Angeles: University of California Press, 1966), 95.
[538] Sandercock, *Cosmopolis II*, 6.
[539] Seng, Eunice, "Politics of Greening: Spatial Constructions of the Public in Singapore," in *Non West Modernist Past*, ed. William Siew Wai Lim and Jiat-Hwee Chang, 143–160 (Singapore: World Scientific, 2011), 149–50, 145.
[540] Ibid.
[541] Corner, "Agency of Mapping," 227.

space.[542] This view is shared by Abidin Kusno, who states that, despite tendencies of planning and policy to dominate and control, local civilians find opportunities for resistance in the "little and "derivative" traditions characteristic of Southeast Asian sociocultural practices.[543] Kusno suggests that architects, such as Tay Kheng Soon, Sumet Jumsai, and Ken Yeang, are among those Southeast Asian practitioners who revived the regionalist discourse in seeking a specific Southeast Asian progressive approach to their urban environments (see Chapter Seven for a discussion on regionalism).[544] It is these kinds of examples, to which Choay alludes, where the utopian model was only ever partially implemented and permitted local opportunities to emerge.[545]

6.5 Concluding Remarks

The utopian vision presented a universal view of the world, and these ideas were translated into visual forms, which included the survey, master plan, and model. These forms of projection may be seen as the standout elements in the transition of mapping from a retrospective role to the projective. The survey was believed to be an accurate description of a geographical reality. Although it was highly rationalized and the geographical information conveyed was through a rigid conceptual frame, the survey was also influenced by the limitations of the techniques available for its production. The master plan became the key tool for organizing large amounts of information about life for a large proportion of people. Sets of social relationships were reduced to rules and principles and presented on a two-dimensional line drawing to inform the production of the model space. The model equates to either a three-dimensional or a two-dimensional reduction of all the utopian thinking, and ultimately represents the living environment that is going to be implemented. The proposed scheme suggests a series of objects or forms that may be universally applied to human settlement. The dilemma for the urban design process is that the field of

---

[542] Ibid.

[543] Ibid.

[544] Abidin Kusno, *Behind the Postcolonial: Architecture, Urban Space and Political Cultures in Indonesia*, ed. Thomas A. Markus and Anthony D. King, Architext Series (London: Routledge, 2000), 58.

[545] See also Yeoh, *Contesting Space*; Mario Gandelsonas, *Urban Text* (Cambridge, MA: Chicago Institute for Architecture & Urbanism (CIAU) / MIT Press, 1991).

operations is so narrow it is unable to connect with the complexities of modern life, thus contributing to the problems of the production of modern urbanism. Some contemporary practitioners are managing to reimagine some creative possibilities within these restrictive design parameters of the modernist mapping paradigms, and their story features next.

CHAPTER SEVEN
Reaction to Modernism

Modern society gained immense benefits because of the Enlightenment being a highly productive period for the development of science and technology. However, its other legacy, in the forms of positivism and rationalism, has been the source of dissatisfaction at many levels, including human environments. A range of responses to modernism have emerged since the mid-twentieth century, including the revisiting of mapping processes in urban design. As Denis Cosgrove suggests, "mapping begets further mappings"; not only are many maps based on their predecessors as a record of comparison, but they are also subject to modification and interpretation once absorbed into the cultural milieu.[546] From within the cultural milieu, first, maps as material objects are able to generate "metaphorical meanings" between, for example, mapping language and ecological visions as well as the psychology of landscape.[547] Second, mappings act as a means of projecting the imagination in the form of "utopias," where a social order is conceived and a spatial order results, represented via the map.[548] Prompted by deep dissatisfactions with modern urbanism and rational urban design strategies,

---

[546] Cosgrove, "Introduction: Mapping Meaning," 14. See also Henri Lefebrve, *The Production of Space* (Oxford, UK: Blackwell, 1991).

[547] Cosgrove, "Introduction: Mapping Meaning," 15. See also de Certeau, *Practice of Everyday Life*.

[548] Cosgrove, "Introduction: Mapping Meaning," 15–16. For discussion of contemporary theoretical utopias, see A. Krista Sykes, *Constructing a New Agenda: Architectural Theory 1993-2009* (New York: Princeton Architectural Press, 2010).

four themes representing some of the major movements that emerged between the second half of the twentieth century and the twenty-first century are examined through the notions of metaphor and utopia. The movements discussed under these themes have developed different mapping approaches to collecting and analyzing social, urban, and environmental data.

The first theme is nomadity and urbanity, under which movements such as the 1960s Situationists fall, as well as projects ranging from New Babylon to the Walking City. Practitioners and theoreticians considered nomadity and urbanity from the utopian perspective, focusing on theorizing the city as an entity in flux. Constantly changing and responding to its inhabitants and global influences, the city is no longer characterized by something fixed and static, typically associated with settlement. The second theme for consideration is landscape and urbanity, which includes the landscape urbanism movement that originated in the 1980s. Landscape urbanism protagonists viewed landscape and urban systems as interacting cohesively and their approach may be considered conceptually linked to metaphorical relations with natural landscapes. However, their process of implementation is closely associated with statistical processes of Enlightenment thinking, which is focused on imaginings of a new order for society, and thus forms part of the utopian genre. Through new mapping processes, landscape urbanists harness large quantities of data and apply them as a design solution. Third, postcolonial theorizations of urban space have been aligned with the theme of regionalism and urban identity. This theme also relates to environmental metaphors, which resonated with practitioners and theorists concerned with the effects of globalization on the distinctive qualities of urban environments. From a theoretical perspective, critical regionalism featured in critical architectural and urban discourse around the 1990s, whereas tropical regionalism and other subcategories of regionalism featured in architectural styles from the same period.[549] The final theme to be examined here is that of ecology, environment, and the city, adopted by the movement known as ecological urbanism, which also has its origins in the 1990s. This movement, across metaphorical associations with ecological factors of landscape, acknowledges that the relationships among

[549] Kenneth Frampton, "Prospects of a Critical Regionalism," *Perspecta* 20 (1983): 147–62.

economics, ecological considerations, and sociopolitical factors come into play to affect the design of the city. This ethos is expressed through mapping media.

Although these four movements surfaced in response to a rationalist context and the early modernist reductive methodology, their urban design approaches maintain, in different ways, modernist undertones. For contemporary practitioners, the difference is that they realize the limitations of modern rationalism, and instead strive to accommodate the non-rational and spontaneous aspects of human life and revisit the imaginative qualities of urban space. However, like their predecessors, their intent remains utopian, whereby they wish to overthrow the modernist paradigm of simplifying urban design to basic geometries. By contrast, under the themes listed below are mappings that display complex factors that contribute to urban design through imaginative thinking, compared with the reductive nature of the early modernists. In addition, mapping technologies and techniques have developed into the twenty-first century to accommodate means of representation to understand our human condition further. Mapping processes and representations demonstrate an integrated relationship between new forms of mapping and recording information based on understandings of the complexities of our world. However, this complexity of thinking does not always translate to a new and improved way of designing. The contemporary map in all of the study cases still exists as a design instrument, not deterministic, but rather informative in its scope.

7.1 Nomadity and Urbanity

Practitioners and theoreticians who considered nomadity and urbanity from the utopian view focused on theorizing the city as an entity in flux, constantly changing and responding to its inhabitants as well as global influences, rather than being characterized by something fixed and static, typically associated with settlement. Movements such as the Situationists considered the city from the perspective of nomadity; their interest in engaging with urban space was by experiencing it through walking or drifting. In their subversion of modernist cartography of the city, the Situationists constructed their own personal spatialities in participating in the *dérive*.[550] One utopian 1960s urban project, New

---

[550] Cosgrove, "Introduction: Mapping Meaning," 16.

Babylon, proposed by Dutch artist Constant Nieuwenhuys, a founding member of the Situationists, involved individuals engaged in projectively mapping their living spaces.[551]

This methodology was intended to overthrow existing societal norms and challenge traditional assumptions and conditions for living and settling. "Settlement," understood within its historic associations of fortification and safety, as well as an urban establishment fixed permanently in a particular position or place, was not desirable for citizens of New Babylon.[552] New Babylon, a utopian model for living, in its proposal of mobility as a radical alternative to urban life, challenged common conceptions of quotidian existence. The connection between temporality, nomadity, and cities also challenges traditional historical interpretations of urban settlements, in that they have been considered, bounded, and physically resolved.

Other examples of the 1960s and 1970s include the following projects: Stanley Tigerman's Instant City project, Buckminster Fuller's Triton project,[553] Archigram's Walking City project,[554] and Mike Webb's Sin Centre project.[555] Nomadity in the contemporary context includes movements, ranging from the physical practice of traversing real space to the capacity of virtual technologies.

Constant's New Babylon originated out of political and social conditions of the 1960s, and followed the Situationists' perception that the modernist movement's negligence was in its failure to recognize the human psychological connection with space.[556] Constant perceived that developments in automation together with population expansion meant increased suburbanization, which led to the demise of creative culture and the destruction of the very life it was meant to support.[557] Believing automation would supersede manual work, Constant imagined free time would be immersed in travel, general ludic behavior, and

---

[551] H. W. F. Saggs, *Everyday Life in Babylonia & Assyria* (London: Jarrold & Sons, 1965), 28.
[552] Simpson and Weiner, *Oxford English Dictionary*, second ed., s.v. "Settlement."
[553] Banham, *Megastructure*, 158.
[554] Ibid., 85. See also Simon Sadler, *Archigram: Architecture without Architecture* (Cambridge, MA: MIT Press, 2005).
[555] Banham, *Megastructure*, 92. See also Christian W. Thomsen, *Visionary Architecture: From Babylon to Virtual Reality*, trans. John William Gabriel (Munich and New York: Prestel, 1994).
[556] "The Great Urbanism Game," Constant Nieuwenhuys' presentation of New Babylon at the Stedelijk Museum in Amsterdam 1960, cited in Wigley, *Constant's New Babylon*, 9.
[557] Ibid, 9.

creating spaces for living. Mapping processes would record engagement in urban spaces of the city. New Babylon was an opportunity for architecture to be used to turn playful urban existence into a new way of living.[558] Constant's New Babylon, as a mobile and temporary city, consists of two discrete components: the aesthetic and the theoretical. The aesthetic component, or utopian model, of New Babylon involves explorations of representational mediums, ranging from mappings to detailed architectural models.[559] The theoretical component, or theoretical model, involves the ideas of unitary urbanism[560] and psychogeography, which originates from Situationist urban philosophy.[561]

The graphic technique used to convey the expansive nature of a *dèrive* in the city is evident in Constant's mappings of Amsterdam. Standard maps are overlaid over the existing historical city center of Amsterdam in a contrasting color.[562] Map fragments represent sectors of the new urban environment that sit over the old city. Constant's graphic portrayals of space were accompanied by textual accounts, lectures to eager architectural students, or manifestoes professing the criteria for the proposed new urbanity, all of which provided insight into the sociocultural qualities of the spaces.[563]

---

558 Ibid, 16. See also Rebecca Solnit, *Wanderlust: A History of Walking* (London: Verso, 2001).
559 Choay, *Rule and the Model*, 10.
560 Wigley, *Constant's New Babylon*, 9.
561 Ibid. See also George Daskalakis, Charles Waldheim, and Jason Young, eds., *Stalking Detroit* (Barcelona: ACTAR, 2001).
562 Wigley, 199, 61. Unitary urbanism involved, according to its authors, the Situationists and Constant, the revival of the "art of construction" in architecture and the practice of the "construction of ambiences" in urban environments. G. E. Debord, "Constant and the Path of Unitary Urbanism" [Original in French, 'Constant et la voie de l'urbanisme unitaire,' written in 1959. Partly published in German in *Constant. Konstruktionen und Modelle* (Essen: Galerie van de Loo, 1960). Translated by Brian Holmes], cited in Wigley, *Constant's New Babylon*, 95. Constant's dissatisfaction with 1960s urban solutions was related to both their myopic nature and the absence of any "socio-cultural" component in the conception or realisation of modern cities. G. E. Debord, "Sur nos moyens et nos perspectives," *Internationale Situationniste* 2, December (1958), cited in Wigley, *Constant's New Babylon*, 95. The practice of unitary urbanism was intended to reinstate cultural practice in the urban realm by focusing upon social contacts and interactions in the intense interior spaces of New Babylon. Thus, unitary urbanism was considered, "the objectification of the creative urge ... the materialisation of dynamic lifestyle. A lifestyle ... which recognises no goal in life, which is not intent on giving life meaning, but which makes life itself the goal, which looks for the fulfilment of this life in daily praxis, a lifestyle, which aims to be the creation of our life." Constant, "Unitary Urbanism" [Original in Dutch. 'Unitair Urbanisme,' lecture, Stedelijk Museum, Amsterdam, December 20, 1960. Translated by Robyn de Jong-Dalziel], cited in Wigley, *Constant's New Babylon*, 132.
563 Cosgrove, "Introduction: Mapping Meaning," 16.

By contrast, the models and the drawings of the built form of New Babylon indicate monolithic structures, occupying and totalizing the existing city. In a similar way to the modernist agenda of overturning the existing modes of production of urban life, New Babylon dominates urban space. Constant blamed utilitarian habits for the production of the existing static and material 1960s urbanism and modernist planning solutions. Only revolution would deliver revised dynamic versions of urban environments.[564] Psychogeography and "Walking the City" were instrumental to overturning the utilitarian habits, through nomadic practice of the *dèrive* or the drift. The subversion of the psychological order of the city (urban planning) was achieved in the drift[565] and resulted in the "construction of atmosphere."[566] The drifter would negotiate the city in a non-systematic way, and this practice was likened to a sporadic, psychological mapping.[567] Unexpected interactions and challenging situations of the detour occurred in residual, marginal urban space of traditional cities.[568]

Spaces of indeterminacy delivered ambient-rich events and atmosphere, and returned to the walker the spontaneity of living. Constant represented these interactions with various modes of mappings. In plans of "New Babylon North", Constant graphically depicts numerous paths and connections of the walker, depicting a sporadic mass of intersecting lines, as the linking device to the mapped sectors in space. This representation attempts to reflect a "rhyzomic" quality to spatial relations.[569]

The model demonstrates the sheer scale of Constant's proposal in three dimensions if it is compared with the cars stationed underneath. The physical space of New Babylon appears to fall short of its theoretical intentions, in that there is a fixity and static quality to the model, which is absent in the two-dimensional mappings. The aesthetic model was an imagined physical long span

---

[564] Wigley, *Constant's New Babylon*, 132.
[565] Ibid., 12.
[566] Ibid., 14.
[567] Iain Chambers, *Migrancy, Culture, Identity* (London: Routledge, 1994), 17. See also Amy Bingaman, Lise Sanders, and Rebecca Zorach, *Embodied Utopias: Gender, Social Change, and the Modern Metropolis* (London: Routledge, 2002). Hannah Arendt, in her book *Human Condition*, argues that indeterminate conditions may also evolve into innovative spaces, provocated through tensions of socioeconomic, political, cultural, religious, and ethnic difference.
[568] Ibid.
[569] See Deleuze and Guattari, *Nomadology*.

structure, stationed more than 30 meters above the existing city fabric and topography, supported on enormous steel pilotis. The ground plane, or existing city level, was set aside for vehicular movement; the first level for the pedestrians and the creation of living spaces; and the top level for playing fields, recreational spaces, and landing space for aircraft.[570] Quotidian spatiality of New Babylon was determined by its fluctuating patterns of use. Inhabitants of New Babylon only ever resided temporarily in any one location. Participators in the interiorized New Babylon space were required, under their own motivations, to create their "patch" for living, sleeping, and playing, through technical manipulation of computer-based atmospheric equipment, movable screens, and partitions. New Babylonians were initially presented with a three-dimensional empty canvas, a voluminous, artificially controlled internal environment where all natural microclimatic systems were excluded. In the absence of external or natural phenomena, interactive technological systems engendered a "sense of place." Sound, air-conditioning, and heating equipment responded to the participator's desired "ambience."[571] The resultant atmosphere was both a psychological and a physical play on spatial experience.

People needed to map their individual experiences of personal and spatial interactions to achieve their own kind of desirable living environment. As a concept, New Babylon played on and borrowed from traditionally organized societies, in which everyday practices motivated urban form, except that mappings were retrospective. The imagined connection between the old and new cities is also displayed in mappings whereby selected traditional cities are remapped by Constant. As an example, a map of Den Haag shows the new sector space in red overlaying the old; this is not a new mode of mapping but simply a reconfiguring and utilization of typical mapping practice. The new mapped representation is reliant on the background context of a traditional map to give the image any meaning. The same technique is used to convey a revised urban layout of Barcelona.

---

[570] Constant, "New Babylon: Outline of a Culture," trans. Paul Hammond, in *New Babylon*, 49–62, (The Hague: Haags Gemeentemuseum, 1974), originally published as "New Babylon, een schets voor een kultuur," rewrite of a chapter from an unpublished book manuscript, written in German between 1960 and 1965, cited in Wigley, *Constant's New Babylon*, 161.
[571] Constant "Unitary Urbanism," 134.

In the context of Françoise Choay's study, and Denis Cosgrove's understandings, New Babylon was ultimately a utopian project.[572] A useful tool to interpret the urban environment and analyze the city, New Babylon in its invention of radical urban situations—utopias—is a crucial conceptual instrument to invoke change.[573] In Choay's terminology, New Babylon's aesthetic and theoretical agendas represent and relate to her "model space" and "model society," respectively, which she identifies as characteristic of the utopian genre. Model space is considered an integral support to the model society, which is opposed to any historically real society.[574] In addition, the model society is disconnected spatially to the traditional city and not subjected to any traditional constraints of time and change.[575] The notion of the model suggests a priority of overturning the existing social order to construct new spatial orders as indicated in the two-dimensional mappings of Constant as well as the three-dimensional proposal of the utopian model.

Constant's concepts have been adopted in contemporary architectural offices such as "Archigram, Architecture Principe, Office of Metropolitan Architecture."[576] Often, in the practice of architecture, architectural projects do not get built, and the ones that do are usually compromised. Therefore, many projects have a utopian aspect, and New Babylon forms part of this genre of the unbuilt.[577] Other favorable aspects depicted by Mark Wigley include Constant's ideas associated with the 1960s cybernetics, which connect New Babylon with contemporary interpretations of cyberspace. In 1958, Constant pioneered a radio-controlled *dèrive*, with the use of the walkie-talkie, to invoke a different type of social space through the "short circuited physical space of the city, fostering the production of a different kind of city."[578]

---

[572] Choay, *Rule and the Model*, 22.
[573] Ibid., 34. See also Helen Rosenau, *The Ideal City in Its Architectural Evolution* (London: Routledge, 1959).
[574] Choay, *Rule and the Model*, 34. For modernist utopian visions, see Nathaniel Coleman, *Utopias and Architecture* (New York: Routledge, 2005).
[575] Choay, *Rule and the Model*, 34. See also Neslihan Turkun Dostoglu, "Architectural Deterministic Thinking in the Development of Urban Utopias, 1848–1947" (PhD thesis, University of Pennsylvania, 1986).
[576] Wigley, *Constant's New Babylon*, 63.
[577] Ibid. See also Stephen Cairns, ed., *Drifting: Architecture and Migrancy*, The Architext Series (London: Routledge, 2004).
[578] Wigley, *Constant's New Babylon*, 65.

The conceptualization of urban space and its relations represented diagrammatically is not a new mode of portraying spatial connections. The main criticism of New Babylon in its representing a type of Situationist architecture, is that there exists a generic contemporary problem of the process of translation from theory to formal realization in the construction of urban spaces. Constant's ideas are relevant in envisaging some lifestyles today, but his proposals operated within a modernist regime. Through Constant's overarching objective of revolutionizing 1960s ways of living and eradicating what was existing then, his utopian concept thrived from the same rationalism and authoritarianism to which he was reacting. There is an attraction in spontaneous situations, in fleeting moments, evoking surprising and unscripted responses as well as creating instantaneous mappings and personal trajectories of experience in space. However, urban environments cannot evolve instantaneously in this sense and then evaporate in the next moment, and by definition, as constructions, they are no longer ephemeral situations. Once Constant's imagined spatialities were mapped into two- and three-dimensional representations, the spontaneity was lost.

Nomadity resonates in contemporary architectural theory because it may occur in the virtual realm, or through modes of thinking, whereby a common link is a quest for new experience as well as altered qualities of human environments. Whether desire for experience is based upon the phenomena of sedentary living, or connected to the psychology of urban environments, the nomad is motivated to wander in search of other events, situations, peoples, and cultures, as well as preferencing mobility itself. Nomadic traversal is "rhyzomic,"[579] and self-directed rather than a conscious movement pattern from place to place, or via delineated paths of urban planners.[580] A nomadic identity is characterized by the ontology of movement in which he or she always exists. A nomad's place is always in flux and therefore has an ephemeral quality. Through his work on conceptualizing nomadity and place experience, Iain Chambers explains that, for the nomad, "the sense of belonging" typically associated with place exists but is constructed in

---

[579] Deleuze and Guattari, *Nomadology*, 6.

[580] Tim Cresswell, "Imagining the Nomad: Mobility and the Postmodern Primitive," in *Space and Social Theory: Interpreting Modernity and Postmodernity*, ed. G. Benko and U. Strohmayer (Oxford: Blackwell, 1997), 364. Place experience in the contemporary world, which may be paralleled with nomadity, has been transformed by modernity and globalisation. Kim Dovey, *Framing Places: Mediating Power in Built Form.* (New York: Routledge, 1999), 48.

mobility.[581] The realm of contemporary citizens/nomads, in their global interconnectedness, has outgrown traditional modes of thinking. An analysis based on psychological and physical shifting relationships with place is required.[582] Contemporary media such as the World Wide Web, virtual "real" space models, and dynamic mapping processes are tools for examining these changing relationships between physical and psychological space. The status of the architect or urban designer has changed since the 1960s. Architectural totalizations, such as New Babylon, equate to art or solely the realm of unrealized utopian solutions surviving in mappings, rather than physical architecture or urban design.

7.2 Landscape and Urbanity

Landscape urbanism protagonists viewed landscape and urban systems as interacting cohesively and conceptually, linked to metaphorical relations with natural landscapes. Modes of mapping employed for realizing landscape urbanism design solutions are closely associated with statistical and thematic maps of Enlightenment thinking. Large quantities of statistical data are analyzed and collated for the use of related themes and metaphor, selectively applied to the design case. Foregrounded by James Corner, Charles Waldheim, and Mohsen Mostafavi, landscape urbanism resulted from postmodernist critiques of modernist urban architectural and planning outcomes. Again, through a search for meaning in human environments, this movement utilized metaphorical readings of the city and the natural represented through mappings to implement utopian ideas into the void spaces of the city. Two contemporary landscape projects are used in the discussion later to demonstrate these mapping processes: the first project is by Rosalea Monacella (2003) and the second design is by Rafael Contreras (2010) from the London school of the Architectural Association.

---

[581] Chambers, *Migrancy, Culture, Identity*, 25. Arjun Appadurai, in his studies of the notions of local and global, suggests that the "ephemeral" quality of a region, in the present era, is due to globalisation. He confirms that "locality itself is a historical product and that the histories through which localities emerge are eventually subject to the dynamics of the global." Arjun Appadurai, *Modernity at Large: Cultural Dimensions of Globalisation* (Minneapolis: University of Minnesota Press, 1996), 1:183.

[582] Cresswell, "Imagining the Nomad," 364.

Early landscape urbanism projects date back to the 1980s with projects such as Parc de la Villette. Once the site of the largest abattoir in Paris, the brief for the competition requiring a new urban park, contained an important requirement—the winning design should embrace possibilities for the twenty-first century.[583] Landscape in this project became representative of a complex transformation of urban infrastructure and a site of public events, while catering for any number of potential configurations for the future. Bernard Tschumi developed a set of relations within the landscape in which to cater for changing and shifting social and urban relationships. Therefore, the intent was that the park be able to cater or engage with the staging of "all sorts of activities, planned and unplanned, imagined and unimagined over time."[584] The intent of such projects is to permit a continuum of changes and events that shape the project, in a process of dynamism. As natural landscapes are always operating within a life-cycle continuum, the collaboration of the natural and the manufactured landscapes of landscape urbanism practices facilitate dynamism not always possible solely within a constructed environment.[585] Although, in practice, Tschumi's sculptures may invite participation and activity that his drawings reflect, they are actually what we have come to understand in contemporary architectural practice as typical built projections. They express geometric concerns of form, scale, and proportional relationships above all else. Although these representations may be producing a different architectural style from previous genres of design, the drawings remain passive in their contribution. A new mode of mapping or translation is not evident.

Waldheim coined the term "landscape urbanism" in a March 1997 conference to define the practices of designers who had replaced landscape as the main mediating form of urban space. This thinking was utopian in its overthrow of the traditional hierarchical relationships of city spaces whereby landscape designs were prioritized over built form.[586] In 1955, Victor Gruen, who was

---

[583] Charles Waldheim, *The Landscape Urbanism Reader* (New York: Princeton Architectural Press, 2006), 38–39, 40.
[584] Ibid., 41.
[585] Ibid. As Waldheim states, "postmodern architects practiced a kind of preemptive cultural regression, designing individual buildings to invoke an absent context, as if neighborly character could contravene a century of industrial economy."
[586] Grahame Shane, "The Emergence of Landscape Urbanism," in Waldheim, *Landscape*

integral to the phenomenon of the megamall, proposed the term "cityscape" in opposition to the notion of "landscape."[587] In this context, landscape was defined as "environment in which nature was predominant."[588] Therefore, a clear distinction between the natural and landscape emerged whereby landscape became understood as an environment that had been altered because of human habitation.[589] Corner identifies four themes in order to demonstrate how contemporary landscape urbanism may be realized in the physical realm. These include "processes over time, the staging of surfaces, the operational or working method, and the imaginary."[590]

> Field diagrams or maps describing the play of those forces are particularly useful instruments in furthering an understanding of urban events and processes. For example, the geographer Walter Christaller's diagrams of population distribution and city planner Ludwig Hilberseimer's diagrams of regional settlement patterns each articulate flows and forces in relations to urban form.[591]

Corner highlights that ecological metaphors are useful in trying to grasp the complexity of contemporary urbanism, and mappings are useful representations through which significant processes in the natural world may be relayed, which enables a complexity of ever-changing relationships to coexist in an environmental system. Linear systems in such cases are deemed obsolete in favor of cyclical and evolving systems responding to ever-changing environmental circumstances. As a legacy of the Enlightenment, landscape

---

Urbanism Reader, 59.

[587] James Corner, "Terra Fluxus", in Waldheim, *Landscape Urbanism Reader* (USA: Princeton Architectural Press, 2006)., 25–26.

[588] Waldheim, *Landscape Urbanism Reader*, 26.

[589] Ibid., 28. Corner believes "the very complexity of the metabolism that drives the contemporary metropolis demands a conflation of professional and institutionalised distinctions into a new synthetic art, a spatio-material practice able to bridge scale and scope with critical insight and imaginative depth."

[590] Ibid. The motivation for landscape urbanism, according to Waldheim, is that "across a range of disciplines, landscape has become a lens through which the contemporary city is represented and a medium through which it is constructed. These sentiments are evident in the emergent notion of 'landscape urbanism.'"

[591] Ibid.

urbanism provides a means of conveying and organizing taxonomies of data. The taxonomies shown in a series of mappings allow the design solution to be rationalized and based on the collated data. As is consistent with projective and contemporary design processes, meanings and identity are sought through metaphors of natural landscapes, whereas maps provide the physical interpretation of these theoretical ideas. James Corner, in his essay "Terra Fluxus," argues that the conceptualization of landscape in today's design schools permits a broader organizational framework within which to study the complexity of issues that affect landscape and, at the same time, a similar framework may be adopted to analyze the contemporary city.[592] The London Architectural Association is one such school that engages with different landscape urbanism strategies.

Ecological, cultural, and design factors are addressed through a series of techniques involving large-scale mapping processes to organize data for strategies from pre-design analysis through to the final design product. Although these are not new ways of mapping, the drawings convey the complexity of information and depth of considerations affecting the design problem. These systems may be seen as stemming from the notion of taxonomies and methods by which to order knowledge, and make sense of it—rationalize it—for the purpose of designing new environments. The design outcomes are also not clear in this particular scheme because the experimentation and focus appear to be the process of mapping.[593] Techniques such as an aerial photograph of the existing site being overlaid by the design reiterate the representational language of Constant, in the superimposition of the new over the old. A similar technique of using the layering of colored modeling clay in specific locations over an aerial view intends to reflect the intensity of programming in some areas of the site. These mapped conceptualizations act as design tools and diagrams to inform the designer for the next stage of design. Innovative modes of representations are aesthetically

[592] James Corner, "Terra Fluxus," in Waldheim, *Landscape Urbanism Reader*, 23. Corner writes, "in particular, thematics of organisation, dynamic interaction, ecology, technique point to a looser, emergent urbanism, more akin to the real complexity of cities and offering an alternative to the rigid mechanisms of centralist planning."

[593] Ibid., 25. Landscape urbanism may be considered a hybrid practice through which to contemplate the realms of landscape and the urban in order to realize design solutions. This gesture is not merely a token green realm within a brownfield site but rather an operational and deliberate strategy, which drives the urban environment to perform to its best capabilities.

interesting, but this does not appear to affect the passive role of the map in the design outcome.

Landscape urbanism concepts and processes of implantation are entirely dependent on maps and mapping processes. The "field" of cartographical association has become a focus in the process of form making in the architectural world, where the notion of surface has sought to combine landscape and the building so roofs and walls become intertwined. For landscape architects and architects, ideas around urban infrastructure suggest future sites for development. The sheer expanse of infrastructure unleashes many sites of potential, or as Corner puts it, "urban infrastructure sows the seeds of future possibility."[594] Grids related to propositions of infrastructure, as well as a means of rationalizing space and information on a map, have an ability to frame objects and topographies within each of their fields, enabling a collation of data within each frame of the grid. The strategy of using mappings and grids to simplify the complexity of information available when studying sites for design interventions also reflects the processes developed in the Enlightenment to reduce space, data, and details of the field study to a manageable, conceivable level and make sense of it all.

Landscape urbanism relies on the ecological aspects of the natural elements in designs to react against Cartesian regularities and uniformities. Although this is achievable for ecological systems, it is more difficult for built form to be as responsive as nature. The natural design components introduce a level of dynamism and freedom in that their growth patterns are not entirely known, and therefore cannot be mapped. Landscape, in projects such as Peter Latz's Duisburg Nord Steelworks Park in Germany, acts as a remedial installation to counter the ill effects of the industrial age. It appears successful at integrating landscape ecologies into large-scale infrastructure projects for the purposes of recreational and environmental regeneration. Data achieved through mapping ecosystems, seasonal variations, populations of species, pedestrian patterns, infrastructure networks, programming of the site and so on eventually require a narrowing of their scope. As highly selective decisive processes to achieve a legible format, mappings maintain their status as instruments rather than direct

---

[594] Corner, "Terra Fluxus," 31. See also Robert R. Churchill, "Urban Cartography and the Mapping of Chicago," *The Geographical Review* 94, no. 1 (2004): 1–22.

realizers or determinants of the design outcome.[595] Although graphically and technologically very sophisticated, and working in three-dimensional formats, these maps do not necessarily stray from the dilemma of the modernist paradigm. This may be partially due to the scope of large-scale environmental and infrastructural systems, which contain an engineering component and result in visually monumental schemes.

Rafael Contreras's "Crackology" explores metaphors and ideas of flows in a non-linear fashion and non-uniform insertions into the everyday fabric of the city, in a variety of models and mappings.[596] His metaphorical premise relates to water filling cracks in a fluid and amorphous manner. Aerial photographs replace the topographical survey as the base of the design process, in their ability to capture the colors and natural attributes of the landscape. Contreras's diagramming indicates how the water encroaches into the cracks of the natural landscape over time, and this symbiotic relationship is mirrored in his design.

This approach concurs with notions of urban morphologists who require within human settlements interaction between activity and spatial patterns, whereby "such linkages, when repeated over time, form islands of local order structuring the larger patterns of global, ecological, and economic flows."[597] Contreras's form making of sinuous buildings on the site, indicating infrastructural and user flows of movement through space, is consistent with these notions of urban morphologies.[598] James Corner notes that landscape via ecology and geography offers an opportunity to attempt new ways of embracing urbanization as it is occurring around the world today, moving it away from its traditional passive role, formerly in movements such as the picturesque.[599]

---

595 See also Shane, "Emergence of Landscape Urbanism," 63.
596 Rafael Contreras, "Crackology," *Kerb 18: PlastiCity FantastiCity: International Design Competition* (2010): 22.
597 Shane, "Emergence of Landscape Urbanism," 63.
598 See also Charles Waldheim, "On Landscape, Ecology and Other Modifiers to Urbanism," *Topos: The International Review of Landscape Architecture and Urban Design* 71, Landscape Urbanism (2010): 21. Charles Waldheim states, "landscape urbanism proposed a critical rereading of the environmental and social aspirations of modernist planning and its most successful models … it proposes a potential recuperation of at least one strand of modernist planning, the one in which landscape offered the medium of urban, economic, and social order."
599 James Corner, "Landscape Urbanism in the Field," in *Topos: European Landscape Magazine,* 71 (Landscape Urbanism), 2010, 26.

Sieve Mapping: Metaphors in Process

Richard Weller is a Perth-based landscape architect and academic. His recent publication *Boomtown 2050* examines future projections of growth for Perth and how these conditions may be addressed within the theoretical confines of landscape urbanism. Weller traces his regionalist approach back to Lewis Mumford, credited with connecting planning to bioregions. This was also the approach of Ian McHarg (1920–2001) in his book *Design with Nature*, which investigated a means of determining sites for development in planning methodology that could be ethically and universally applied.[600]

The landscape provided clues to the optimal conditions for development. Establishing habitat corridors and other natural site attributes to obtain one holistic map in turn identified certain locations for "suitable" development. This process was called "sieve mapping" whereby certain regions would be deleted from the development site possibilities based on inadequate or inappropriate conditions. Large tracts of land would be rendered unsatisfactory for development. The limitations of this method included sorting through large quantities of data associated with complexities of contemporary development, which far exceeded the premise and parameters of McHarg's study.[601]

Despite the procedural limitations, Weller's proposal in *Perth 2050* identifies five different options, under the following thematics: Food City, POD (Performance Oriented Development) City, Surf City, Sky City, River City (refer Figure 16).[602] The thematic approach is intended to inspire a diversity of development and a variety of ways of thinking about food production, transport, typologies, and densities for housing. Within the genre of traditional landscape urbanism approaches, mappings convey data ranging from infrastructure extents to areas required for food production. These mappings are accompanied by textual accounts explaining the assumptions and logic behind the statistics. Maps aid the method of design, in a way that "the landscape thus forms the foundation from which the various urban-growth scenarios emerge."[603] Modes of representation,

---

[600] Richard Weller, "Planning by Design: Landscape Architectural Scenarios for a Rapidly Growing City," *Journal of Landscape Architecture* Autumn (2008): 6–17, 167.
[601] Ibid., 171.
[602] Ibid., 11.
[603] Ibid., 165.

such as fly-over computer-modeled projections, attempt to place these PODs in a geographical context to suggest spatial and physical relationships between the sectors for development (refer Figure 17). Paradoxically, these representations, at the scale and viewing angle of the proposal, convey spatial isolation and disconnection between sectors. Problems associated with thematic mappings have their origins in the Enlightenment and, although mapping graphics are sophisticated, the outcome highlights the inadequacies of this type of representation.

Cosgrove traces the origins of statistical mapping in Europe to the late eighteenth and early nineteenth centuries, graphically relayed through charts, graphs, and maps.[604] Thematic mappings also featured examples such as Alexander von Humboldt's meteorological map showing weather patterns rather than solely statistical information. Thematic maps were popularized around the mid-nineteenth century in their demonstration of medico-statistical information for social planning.[605] The authority of these maps is based on their statistical foregrounding; however, two main areas of critique include that the spatial propositions put forward by these projections are often vague and that they incorporate modes of representation highlighting different areas with shading. Cosgrove suggests that shading in dark tones carries certain "moral connotations" of negative attributes of the space, such as emphasizing regions of socioeconomic decline or areas on the map where disease predominates.[606]

7.3 Regionalism and Urban Identity

Postcolonial theorizations of urban space typically analyze the theme of regionalism and urban identity. Regionalism, as a theoretical concept, references historical and traditional materials, traditional archetypes, climatic factors, and ways of seeing the world, to carry through some associations of sociocultural context into the production of human settlement. In visual terms, the picturesque movement captured a particular way of representing a traditional context, which

---

604 Denis Cosgrove, *Geography and Vision: Seeing, Imagining and Representing the World* (London: I. B. Tauris, 2010), 162.
605 Ibid., 163.
606 Ibid., 164–5. See also John K. Wright, 1947. "Terrae Incognitae: The Place of Imagination in Geography," *Annals of the Association of American Geographers* 37 (1947): 1–15.

was later seen as highly contrived and orientalist in its romanticism of the native. Regionalism has also suffered from criticism that it romanticizes the past in seeking reasons for designing and representing a region, when those references no longer have the same meanings for contemporary space.

Contemporary forms of tropical regionalism use environmental metaphors for motivations to design and represent up-to-date interpretations of region, which resonated with practitioners and theorists concerned with the effects of globalization on the distinctive qualities of urban environments. In seeking cultural identity within a context, this movement attempts to secure the concept of regionalism in its built environments in its utilization of maps and architectural elements as material objects to relay a language and style constructed about a place.

Similar to picturesque graphic interpretations of particular contexts, drawings stemming from architects' work with a regionalism agenda create ambiguous connections with the settlement, environment, and culture they are attempting to reference. Despite the written rhetoric with, for example, Ken Yeang's "Green skyscraper" sketched as a "theory of vertical urban design," permits a graphic impression of an architect's conceptual drawing whilst any regionalist associations are not explicit.[607] Yeang sees the skyscraper as a "vertical city" containing pedestrian routes, public realms of greened zones, civic areas, some floors having higher densities to others, and social space also inherent in the design.[608] Although valid architectural concepts, these buildings and their representations do not reflect solely Southeast Asian ideals. Environmental sustainability and ecological considerations of design have a global audience. A constructed spatiality through the employment of elements seen as pertaining to that region, addressing particular socio-climatic conditions, only implies an association with that context. These elements may indeed be transferrable to other contexts of similar geographies and climates and just as easily become synonymous with that place. Regionalism also proposes a utopian notion that

---

[607] Robert Powell, *Rethinking the Skyscraper: The Complete Architecture of Ken Yeang* (London: Thames and Hudson, 1999), 159.
[608] Ibid., 156. See also Ken Yeang, "The Skyscraper Bioclimatically Considered," in *Lecture Series Accompanied by an Exhibition of T. R Hamzah and Yeang's Architecture* (New York: American Institute of Architects, 1997).

there might exist some type of regional ideal or authentic form to suit both context and peoples of a certain place in the world. This search for meaning of the projective designer is based on a nominated space demarcated on a map and is problematic, yet symptomatic, of societies that no longer identify with sociocultural, spiritual, or religious connections to place. Even in premodernity, aspects of complexity in everyday lives of peoples extended beyond environment and climate. A quest for meaning based on climatic and environmental factors dismisses the value of other everyday activities, which also contribute to a holistic picture of peoples' lives.

In his essay "The Concept of Regionalism," Alan Colquhoun questions the rationale behind the regionalist position in contemporary societies, when particular sociocultural practices are not necessarily related or tied to a specific geographic region.[609] Colquhoun identifies the origin of regionalism in the separation of romanticist and rationalist traditions. Nativist and naturalist approaches of the eighteenth and nineteenth centuries subsequently were revived in discourses contemplating built space.[610] Based on a utopian notion of a regional ideal, both from a locational and social perspective,

> according to this model, all societies contain a core, or essence, that must be discovered and preserved. One aspect of this essence lies in local geography, climate, and customs, involving the use and transformation of local, "natural" materials.[611]

---

[609] Alan Colquhoun, "The Concept of Regionalism," in *Postcolonial Space(S)*, eds. Gulsum Baydar Nalbantoğlu and Chong Thai Wong, 13–24 (New York: Princeton Architectural Press, 1997), 13.

[610] Ibid., 14. Postcolonial spatiality discourse of the late 1990s stems from the colossal oppositions set up in the eighteenth century between "East/West, traditional/modern, natural/cultural, structural/ornamental." This discourse, according to editors of the book *Postcolonial Space(s)*, edited by Gulsum Baydar Nalbantoğlu and Chong Thai Wong, has been given a contemporary voice in regionalist arguments of the present day. Nalbantoğlu and Wong, *Postcolonial Space(S)*, 8. Regionalism is defined in Colquhoun's essay as "a search for an authentic cultural and architectural essence ... formulated 'precisely at the moment when the phenomena that it described seemed to be threatened and about to disappear.'" An argument of authenticity is controversial as it forces a reductive view of whatever is supposed to be authentic, given that representations of authentic things result from later analyses.

[611] Colquhoun, "The Concept of Regionalism," 14. Theories of the late nineteenth century relating to the rationalisation of social life under industrial capitalism are linked to concepts of regionalism in that there was a general disenchantment due to the effects of "iron age" of capitalism, from Max Weber's analysis.

Concepts of regionalism and what is understood as a region are central to what is established on a map. The motivations around this utopian model date back to the eighteenth century when all the attributes of indigenous and local societies were perceived as threatened or highly compromised. In relation to architecture or built form, the connection to natural or raw materials was important, if a regional quality to the buildings were to be achieved. However, this connection is more of a "desire," as Colquhoun puts it, rather than any particular attempt at authenticity.[612] However, in seeking a mandate of authenticity, the regionalist doctrine in effect invokes an oversimplification of what is a very complex cultural form of production.

Colquhoun believes that, compared with traditional cultures, modernity permits a level of standardization not possible in traditional practices of production. By contrast, the dissemination process is infinitely more variable in the number of different forms of communications on offer in the contemporary environment, compared with traditional scenarios. These transitions in forms of production, with the resultant multivalent nature of coding available in the modern world, have undermined what was previously a much more rigid process that linked traditional practices to a region.[613]

> The concept of regionality depends on it being possible to correlate cultural codes with geographic regions. It is based on traditional systems of communication in which climate, geography, craft traditions, and religions are absolutely determining. These determinants are rapidly disappearing and in large parts of the world no longer exist. That being the case, how is "value" established?[614]

---

[612] Ibid., 22.

[613] Ibid.

[614] Ibid. See also Kusno, "Imagining Regionalism." Abidin Kusno writes, "the shaping of the built environment is also a writing of the history of a nation." Kusno, as is the case for other postcolonial studies' authors, focuses on the colonized and their agency to resist being dominated by the colonizers, and their role is shaping their urban environments. Other authors, such as Brenda Yeoh in her scholarly inquiry, "Imagining Regionalism," into the shaping of modern Singapore, reveal the strength and abilities of the colonized to substantially affect the evolution of their urban spaces. The city of Singapore unfolds as a result of activities between both British and Chinese inhabitants. The value of these understandings may be seen through processes of

Mapping in premodernity represented, and was closely associated with, a physical and often ephemeral context, as were other modes of cultural production. In an increasingly globalized world, the notion of regionalism as an explicit form of expression in maps and architecture, oversimplifies spaces of modernity. For the case of mapping, maps are not tied stylistically, technologically, or socioculturally to their place of production. Their physical production follows cartographical standards and codes where the point of difference is in interpretation or the knowledge base of the viewer. The map reader, based on his or her sociocultural or sociopolitical position, responds to the subtleties of meaning of the mapped projection. However, overall, literal regionalism references have lost their meaning in a global environment. Henri Lefebvre argues that the knowledge base required to decode the infinite number of codes and meanings associated with social spaces of modernity is an unknown entity.[615] Lefebvre states, "we are confronted not by one social space but by many—indeed, by an unlimited multiplicity or unaccountable set of social spaces."[616] However, this aspect of mapping often is overlooked because maps in their artisanship and presentation are often very persuasive, and unlike with writing, it is not possible for mapmakers to express their limitations. Therefore, maps have a tendency to be viewed as "naïve representations of reality." On the contrary, map knowledge is culturally reproduced, so mapping codes and skills have to be learned so that we recognize the reality with which we are being represented in maps.[617]

In essence, maps represent a view of the world, constructed by the mapmaker, and these selective interests are further revealed and built upon through the process of design.[618] Ecological urbanism adopts a variety of

---

mappings in that, once the colonizers reached the shores of Southeast Asia, typically, representations of space were largely influenced or commissioned under colonial instruction. Kusno states "the ways in which 'old' colonial imaginings—found to be useful in the memories of 'modern' postcolonial governing elite—continue to surface and to structure the politics of time and space in the later postcolonial regime." Kusno, "Imagining Regionalism," 19. Kusno believes the shaping and values of contemporary architectural and urban space are focused on sociopolitical motivations rather than sociocultural concerns.

[615] Lefebrve, *Production of Space*, 85.

[616] Ibid.

[617] Pickles, *History of Spaces*, 35, 61.

[618] Wood, *Power of Maps*, 183.

mapmaking principles to serve its interests in shifting the focus on the urban to ecological systems, as a means of realizing a sustainable urban future.

7.4 Ecology, Environment, and the City

Themes linking built form with ecology, flagged in the book *Ecological Urbanism*, deliver a series of different approaches to contemporary design, reviving the imaginative qualities and complexities of mapping. The predominant modes of mapping concur with methods outlined above, using statistics and metaphor to arrive at design proposals. According to the author, Mohsen Mostafavi, the intent is to offer "a framework that through the conjoining of ecology and urbanism can provide the knowledge, methods and clues of what the urban can be in the years to come."[619] Therefore, Mostafavi envisages an urbanism where there are opportunities for the coexistence of ecology and urban rather than their current positioning in opposition to one another. Because of our contemporary challenges of rapid urbanization as well as declining natural resources, a reconsideration of design on a large-scale is required.[620] Mostafavi is thus proposing design collaborations globally to address environmental and urban problems that are unprecedented. The task consists of confronting a complex array of considerations ranging from ecological, economic, social, political, and cultural, all of which affect local and global communities.

Drawing influence from a 1980s philosophy written by Félix Guattari in "The Three Ecologies," Mostafavi references Guattari's idea of "ecosophy." This concept demonstrates the holistic relations between environment, social relations, and human subjectivity.[621] Pressing for revolution of a global scale from social and cultural perspectives, Guattari's key aim was to redefine methods of "the production of material and immaterial assets."[622] Guattari recognizes the importance of both individual and group actions to realize his aims of revolution and ultimately change the production of urban life.[623] Mostafavi also promotes

---

[619] Ibid., 32.
[620] Mohsen Mostafavi, "Why Ecological Urbanism? Why Now?" *Topos: The International Review of Landscape Architecture and Urban Design* 71, Landscape Urbanism (2010): 31.
[621] Ibid., 34.
[622] Mostafavi, 2010, 35.
[623] Mostafavi, "Why Ecological Urbanism?" 35. Mostafavai states, "Extending Guattari's suggestions that the 'ecosophic problematic' has the capacity to define a new form of human

radical change in human activity and patterns of living for the reorganization and rethinking of a sustainable existence. The magnitude of the global environmental predicament and the need to act, Mostafavi believes, requires a revision of urban design methodologies entrenched in master planning processes. As the projects of ecological urbanism address large-scale problems, this exposes the limitations of traditional planning and urban design tools of production. Cosgrove states:

> Urban space and cartographic space remain inseparable. As each is transformed the relationship between them alters, and current visual technologies mean that the opportunity for creativity in shaping and recording urban experience is greater than ever, as too is the need for making and meaning of both public and private urban spaces.[624]

The reality of the ecological urbanism projects as presented in Mostafavi's book, although harnessing the visual technologies to which Cosgrove refers, is that the majority of the projects utilize diagramming as well as innovative presentations of textual reports in representing their design ideas and methods. One such analysis is Richard Forman's book chapter and method identified as "spatial principles for nature in urban regions."[625] Forman's diagramming references taxonomies of the Enlightenment, and serves as principles through which to realize his concept of urban ecology. Based on four main categories, from natural to built form, they propose ecological alterations based on the actions of humans decreasing "natural vertical structure, horizontal pattern, and/or flows and movements."[626] According to Forman, areas of degradation can be improved by ecological planning, limiting suburban sprawl, which seems to

---

existence, we might consider the impact of the ecological paradigm not only on ourselves and our social actions in relation to the environment, but also on the very methods of thinking that we apply to the development of disciplines that provide the frameworks for shaping those environments."

624 Cosgrove, *Geography and Vision*, 182.

625 Richard T. T. Forman, "Urban Ecology and the Arrangement of Nature in Urban Regions," in *Ecological Urbanism*, eds. Mohsen Mostafavi and Gareth Doherty, 312–323 (Baden, Switzerland: Lars Muller, 2009), 313.

626 Ibid., 314. See also Peter Newman and Isabella Jennings, *Cities as Sustainable Ecosystems* (Washington, DC: Island Press, 2008).

involve a mix of fine-grain and coarse-grain landscape interventions.[627] Forman's mappings form part of his focus on a combination approach to urban planning and ecological science. His evaluation of thirty-eight regions around the world establishes patterns of natural systems in contact with human settlements.[628] These modes of representation and research do not always indicate physical proposals, but reflect conceptual urban design frameworks. Although demonstrating complex sets of relations, they have not really strayed aesthetically and functionally from the modernist utopian model.

Koolhaas believes the future of ecological urbanism lies in the innovation of large-scale engineering projects that embrace renewable energy sources to bring about a change in the way we construct our urban landscapes. OMA's North Sea Project is an example of the attempt by countries bordering the North Sea, such as Norway, Sweden, Denmark, Holland, Belgium, and England, to form an amalgamation of political force to harness wind energy from the North Sea for power generation to these countries.

The initiative is also intended to utilize other energy sources, such as solar, tidal and wind, to form a European energy grid.[629] This project has a different spatiality from conventional or traditional infrastructure projects. Representations of the scheme highlight components of the project that will be utilized to produce energy. These components fluctuate because they are dynamic elements of the ecological environment, that is, wind, tidal performance, and solar aspects. The North Sea project bridges countries and their planning systems to bring about a far greater level of coordination to benefit participatory countries. This holistic initiative for the production of energy may be instrumental in driving future thinking about the formation of settlements. Modifying behaviors to regulate consumption of energy is one of the global challenges affecting conceptions of future human environments.

---

[627] Forman, "Urban Ecology," 316. See also Rem Koolhaas, "Advancement versus Apocalypse," in Mostafavi and Doherty, *Ecological Urbanism*, 58.

[628] Forman, "Urban Ecology," 318. See also William McDonough and Michael Braungart, *Cradle to Cradle—Remaking the Way We Make Things* (New York: North Point Press, 2002).

[629] Koolhaas, "Advancement versus Apocalypse," 70–71. See also J. Howard Kunstler, *The Long Emergency* (New York: Atlantic Monthly Press, 2005); E. Beyer, A. Hagemann, and T. Rieniets, *Atlas of Shrinking Cities* (Ostfildern, Germany: Hatje Cantz, 2006).

7.5 Concluding Remarks

The different approaches discussed in this chapter operate within the rationalist frame of the modernist paradigm. This frame persists as a legacy of the Enlightenment, and although it has permitted, in postmodernity, innovative thinking and conceptualizing of urban environments, mappings have not responded to the active potentials flagged by geographical theorists such as Corner and Cosgrove. Other than design drawings, the map's role in the realization of spaces of landscape urbanism, regionalism, and ecological urbanism remains a passive tool. Perhaps the one exception is philosophy, which enacted mapping via the practice of psychogeography, yet these actions did not result in the creation of built form or new space. However, Cosgrove continues to have faith that the visual and creative technologies available to contemporary designers today will aid a shift in mapping practice, and move it beyond notions of metaphors and utopias. Virtual computer tools will realize new methods of activation in mapping and potential for altering what we have come to associate as the reductionist map.[630]

---

[630] Cosgrove, *Geography and Vision*, 182.

CONCLUSION

As an urban designer, I have presented a study on the story of mapping based on a search to discover the modern map's origins, to further understand its limitations as well as its possibilities for urban design practices in Southeast Asian and Western societies. This story has uncovered three main modes of mapping: the imaginative and non-rational modes of premodernity; the rational, literal, and reductive modes of early modernity and the Enlightenment; and the return of the imaginative in postmodernity in an attempt to redeem some of the lost qualities in processes of modernization. These three forms of mappings demonstrated particular projections: premodern maps were characterized by their non-orthographic and non-perspectival but illustrative nature; Enlightenment ideas were expressed in the reductive forms of the exact science, expressed in the utopia, the survey, and the master plan; and postmodern practices have revisited projections of early modernity to introduce statistical, thematic, and layering techniques to these earlier forms to recapture the imaginative potentials of mapping.

The key imaginative qualities of maps of premodern times were in their adoption of subjects and the projection of ideas about places, from the mythical to the sacred perspectives. For Southeast Asian premodern peoples, from what we know, mappings appear to have served purposes of sociocultural and religious instruction on earthly and non-earthly relations, as well as embracing mythical ideas. Concepts of this scope were largely tied to retrospective representations of settlement or inscriptions on major temples and sites of worship. Thus, premodern mappings do not appear, from the evidence available to us, to have informed the process of settlement creation in a highly preconceived and projective way in the Southeast Asian region. From the evidence depicted in the primary sources, representations of built form extended to retrospective maps of religious temples and shrines as well as military maps of settlements produced for purposes of conquering neighboring territories.

By contrast, the early modernist period and the European Enlightenment signified a period of transitional mapping practices, from declining preoccupations with spiritual and mythical space to quests for a precise geography. Geography was the core science that changed understandings of the earth and how we describe these understandings in modern maps. Requirements for an increasingly accurate map of the world corresponded with the Second Great Age of Exploration. Geographical knowledge of the New World was refined by firsthand accounts and with objectives of securing and demarcating territories for empire expansion. Therefore, the survey emerged as a critical instrument for negotiating issues of land ownership as well as the information pertaining to topographical territory characteristics, and a means by which to chart natural resources and commodities of the region. The other two forms of mapping related to colonial expansion were utopian representations and the master plan. Based on Francoise Choay's work, we learned that conceptualizing urban space in writing originated as part of Western discourse on human settlement. Out of this genre of writings emerged the notion of utopia, whereby sets of rules and principles, foregrounded in critical understanding of urban environments, could all be rationalized to a theoretical frame for building. The master plan contributed to the representation of the model space in its further reduction of utopian ideals into a two-dimensional drawing.

These three legacies of Enlightenment geography—the survey, utopian model, and master plan—led to the remapping of territories in colonial Southeast Asia. A shift in mapping practices occurred whereby retrospective approaches encompassing imaginational spatialities of the sacred and profane were disregarded. Projective techniques utilizing the survey, utopian model, and master plan were prioritized, and a universality of the mode of production of human environments followed. The Industrial Revolution and the machine age advanced the perceived authority of rationalization. Because rationalization in the factory related to efficiencies of time and costs while maximizing mass production, this thinking motivated the modern movement's embrace of rational forms of planning.

A reaction to modernism, although still within the legacy of a rational framework, seen in postmodern practices has tried to reconnect with the imaginational aspects of mapping practices. From the Situationists in Europe to tropical regionalism in Southeast Asia, these movements have strived to reject modernist paradigms and reimagine new urbanisms. Statistical and metaphorical methods of landscape and ecological urbanism movements have also attempted to create more responsive and engaging human environments. However, ultimately, the map's role has remained passive in the mode of production of settlement.

Along this journey of inquiry, I have reconnected with lost meanings and functions of mapping to reconcile the loss of identity and quality in some urban designed spaces of postmodernity. The future possibilities of mapping seem to be in not how they look but what they do and how they function. To borrow the words of David Harvey, opportunities for the role of mapping in the design of urban environments may be in the shifting of maps from being passive tools to being active interfaces, and thus focusing on a "utopia of process" over a "utopia of form."[631] Ecological urbanism prioritizes mapping process, and goes some way to highlighting the significance of interrelated systems in a sustainable world. This movement invites further research in that it appears to also facilitate formal qualities of design in responding to the changing nature and needs of people, plants, and animals, all of which are essential to the production of fast-paced urban environments.

---

[631] Harvey, *Spaces of Hope*.

BIBLIOGRAPHY

Abu-Lughod, Janet. 1991. *Before European Hegemony: The World System A.D. 1250-1350*. Oxford, UK: Oxford University Press.

Acker, William Reynolds Beal. 1954-1974. *Some T'ang and Pre Tang Texts on Chinese Painting*. Translated by William Reynolds Beal Acker. 2 vols. Vol. 2.1 Leiden: E. J. Brill.

Aczel, Amir D. 2001. *The Riddle of the Compass: The Invention That Changed The World*. New York: Houghton Mifflin Harcourt.

Ahmad, S Maqbul. 2005. "Kharita." In *The Encyclopedia of Islam: New Edition*, edited by C E Bosworth and et al. Brill Academic Publishers.

Akbar, Jamel. Khatta and the Territorial Structures of Early Muslim Towns. Dammam, Saudi Arabia: King Faisal University.

———. 1988. *Crisis in the Built Environment: The Case of the Muslim City*. Singapore: Concept Media Pty Ltd.

Akkach, Samer. 2002. "Religious Mapping and the Spatiality of Difference." *Thresholds* no. 25 (Sacrosanct):69-75.

———. 2005. *Cosmology and Architecture in Premodern Islam: An Architectural Reading of Mystical Ideas*. Edited by Seyyed Hosein Nasr, *SUNY series on Islam*. Albany, USA: State University of New York Press.

Al Sayyad, Nezar. 1992. *Forms of Dominance: On the Architecture and Urbanism of the Colonial Enterprise*. Brookfield, VT: Avebury.

Al Sayyad, Nezar, and Mrinalini Rajagopalan. 2009. "Colonial City." In *Encyclopedia of Urban Studies*, edited by Ray Hutchison, 166-71. Thousand Oaks, California: SAGE Publications Inc.

Alberti, Leon Battista 1404-1472. *On the art of building in ten books*. Translated by Joseph Rykwert, Neil Leach and Robert Tavernor. 1988 ed. Cambridge, Mass: MIT Press.

Alfonsin, Anthony. 2002. *The Struggle For Modernism: Architecture, Landscape Architecture and City Planning at Harvard*. New York: W. W. Norton.

Allen, Jane. 1991. "Trade and Site Distribution in Early Historic Period Kedah: Geoarcheological, Historic and Locational Evidence." *Indo-Pacific*

Prehistory 1990 no. 1 (Bulletin of the Indo-Pacific Prehistory Association No. 10. Indo-Pacific Prehistory Association, Canberra Australia):307-319.

Alpers, S. 1983. *The Art of Describing: Dutch Art in the Seventeenth Century*. Chicago: The University of Chicago Press.

Amin, Ash, and Nigel Thrift. 1992. "Neo-Marshallian nodes in global networks." *International Journal of Urban and Regional Research* no. 16 (4):571-587.

Anderson, Benedict 1991. *Imagined Communities: Reflections on the origin and spread of nationalism*. Revised second edition ed. London New York: Verso.

Anderson, Stanford, and Martha D. Pollak. 1997. *The Education of the Architect: Historiography, Urbanism and the Growth of Architectural Knowledge*. Cambridge, Massachussetts: MIT Press.

Andreotti, Libero , and Xavier Costa. 1996. *Theory of the Dèrive and other Situationist writings on the city* Barcelona : Museu d'Art Contemporani de Barcelona: Distribution, ACTAR.

Appadurai, Arjun. 1996. *Modernity at Large: Cultural Dimensions of Globalisation*. Vol. 1. Minneapolis: University of Minnesota Press.

Armstrong, W, and Terry McGee. 1985. *Theatres of accumulation: Studies in Asian and Latin American Urbanisation*. London and New York: Methuen.

Arnheim, Rudolph. 1969. *Visual Thinking*. California: University of California Press.

Askew, Marc, and William S. Logan. 1994. *Cultural Identity and Urban Change in Southeast Asia: Interpretative Essays*. Geelong, Victoria, Australia: Deakin University Press.

Atkin, Tony, and Joseph Rykwert. 2005. *Structure and Meaning in Human Settlements*. Philadelphia, Pennsylvania: University of Pennsylvania Museum of Archaeology and Anthropology.

Atkinson, T D. 1955. Points of the Compass. In *Encyclopedia of Religion and Ethics*. New York.

Augae, Marc. 1995. *Non-places: Introduction to an Anthropology of Supermodernity*. Translated by John Howe. New York: Verso.

———. 1999. *An Anthropology for Contemporaneous Worlds*. Translated by Amy Jacobs, *Mestizo Spaces*. Stanford, California: Stanford University Press.

Auge, Marc, and Tom Conley. 2002. *In The Metro*. Minneapolis: University of Minnesota Press.

Bacon, Edmund. 1967. *Design of Cities*. London: Thames and Hudson.

Bacus, Elizabeth, Ian Glover, and Peter Sharrock. 2008. *Interpreting Southeast Asia's Past: Monuments, Image and Text*. Singapore: NUS Press.

Bacus, Elizabeth, Vincent Piggot, and Ian Glover. 2006. *Uncovering Southeast Asia's Past: Selected Papers from the 10th International Conference of the European Association of Southeast Asian Archaeologists*. Singapore: NUS Press.

Banham, Reyner. 1976. *Megastructure:Urban Futures of the Recent Past*. London: Thames and Hudson Ltd.

Barnet, Richard J, and Ronald E Muller. 1974. *Global Reach: The power of multinational corporations*. New York: Simon & Schuster.

Baudrillard, Jean. 1994. *Simulcra and Simulation*. Translated by Sheila Faria Glaser. Ann Arbor: The University of Michigan Press.

—. 2006 *Utopia deferred : writings from Utopie, (1967-1978)* Translated by Stuart Kendall. New York Semiotext(e).

Bellwood, Peter. 1997. *Prehistory of the Indo-Malaysian Archipelago*. Honolulu: University of Hawai'i Press Original edition, 1985. Reprint, 1997.

Benedikt, Michael. 2005. Cityspace, Cyberspace, and the Spatiology of Information. In *New Urbanism Symposium*. Austin, Texas: The University of Texas

Benhabib, Seyla. 2002. *The Claims of Culture: Equality and Diversity in the Global Era*. Princeton, N. J: Princeton University Press.

Benjamin, Walter, Rolf Tiedmann, Howard Eiland, and Kevin McLaughlin. 1999. *The Arcades Project / Walter Benjamin*. Translated by Howard Eiland and Kevin McLaughlin. Cambridge, Massachussetts: Belknap Press.

Benko, Georges, and Ulf Strohmayer. 1997. *Space and Social Theory: Interpreting Modernity and Postmodernity*. Oxford, UK: Blackwell Publishers Inc.

Bennett, Steven J. 1978. "Patterns of the Sky and Earth: A Chinese Science of Applied Cosmology." *Chinese Science* no. 3:1-26.

Bermingham, Ann. 1986. *Landscape and Ideology: The English Rustic Tradition, 1740-1860*. Berkeley, California: University of California Press.

Betteridge, Thomas. 2007. *Borders and Travellers in Early Modern Europe*. Aldershot, Hampshire, England: Ashgate Publishing House.

Beyer, E, A Hagemann, and T Rieniets. 2006. *Atlas of Shrinking Cities*. : Hatje Cantz Publishers.

Bingaman, Amy, Lise Sanders, and Rebecca Zorach. 2002. *Embodied utopias : gender, social change, and the modern metropolis* London: Routledge.

Bishop, Ryan, John Phillips, and Wei-Wei Yeo. 2003. *Postcolonial Urbanism: Southeast Asian Cities and Global Processes*. New York: Routledge.

Black, Jeremy. 2000. *Maps and History:Constructing Images of the Past*. New Haven, London: Yale University Press.

Boisselier, Jean. 1976. *Thai Painting*. 1st ed: Kodansha America, Inc.

Bonfante, Larissa. 2005. "Etruscan Boundaries and Prophecy." In *Structure and Meaning in Human Settlement*, edited by Tony Atkin and Joseph Rykwert, 153-163. Philadelphia: University of Pennsylvania Museum of Archaeology and Anthropology.

Boreden, Iain, and Sandy McCreaery. 2001. *New Babylonians*. Vol. 71, *Architectural Design*. London: John Wiley & Sons.

Borges, Jorge Luis. 1964. "Museum: On Rigor in Science." In *Dreamtigers*, edited by Mildred Boyer and Harold Morland. Austin: University of Texas Press.

Bosselmann, Peter. 1998. *Representation of Places: Reality and Realism in City Design*. USA: University of California Press.

Bowen, Margarita. 1981. *Empiricism and Geographical Thought From Francis Bacon to Alexander Humboldt*. Cambridge: Cambridge University Press.

Boyarsky, Nicholas, and Peter Lang. 2003. *Urban Flashes Asia: New Architecture and Urbanism in Asia*. September/October 2003 ed. 5 vols. Vol. 73, *Architectural Design*. London: John Wiley & Sons.

Boyer, M. Christine. 1994. *The City of Collective Memory: Its Historical Imagery and Architectural Entertainments*. Massachussetts: MIT Press.

———. 1994. *The City of Collective Memory: Its Historical Imagery and Architectural Entertainments*. Cambridge, MA: MIT Press.

———. 1996. *CyberCities: Visual Perception in the Age of Electronic Communication*. New York: Princeton Architectural Press.

Brand, Diane. 2004. "Surveys and Sketches: 19th-century Approaches to Colonial Urban Design." *Journal of Urban Design* no. 9 (2):153-75.

Broeze, Frank. 1989. *Brides Of The Sea:Port Cities of Asia From the 16th-20th Centuries*. Kensington.

———. 1997. *Gateways of Asia: Port Cities of Asia in the 13th-20th Centuries*. London.

Bronson, Bennet. 1977. "Exchange at the Upstream and Downstream Ends: Notes Toward a Functional Model of the Coastal States in Southeast Asia." In *Economic Exchange and Social Interaction in Southeast Asia: Perspectives From Prehistory, History and Ethnography*, edited by K L Hutterer. Ann Arbor.

Brown, Robert L. 1988. "Bodhgaya and South-east Asia." In *Bodhgaya: The Site of Enlightenment*, edited by Janice Leoshko, 101-24. Bombay: Marg Publications.

Buisseret, David. 1998. *Envisioning the City: Six Studies in Urban Cartography*. Edited by James Akerman. Chicago: The University of Chicago Press.

Bunnell, Tim. 2002. "Cities For Nations? Examining the City-Nation-State Relation in Information Age Malaysia." *International Journal of Urban and Regional Research* no. 26 (2):284-298.

Burke, P. 1969. *The Renaissance Sense of the Past*. London: Edward Arnold.

Butlin, Robin A. 1992. "Ideological Contexts and the Reconstruction of Biblical Landscapes in the Seventeenth and Early Eighteenth Centuries: Dr Edward Wells and the Historical of Geography of the Holy Land." In *Ideology and Landscape in Historical Perspective*, edited by Alan R. H. Baker and Gideon Biger. Cambridge: Cambridge University Press.

Cahill, James. 1982. *The Compelling Image: Nature and Style in Seventeenth-Century Chinese Painting*. Cambridge: Harvard University Press.

Cahill, James F. 1960. "Confucian Elements in the Theory of Painting." In *The Confucian Persuasion*, edited by Arthur F. Wright. Stanford: Stanford University Press.

Caird, G B. 1988. *The langugage and imagery of the bible*. London: Duckworth.

Cairns, Stephen. 2004. *Drifting: Architecture and Migrancy*. Edited by Thomas A. Markus and Anthony D. King, *The Architext Series*. London: Routledge.

Campbell, Tony, and Marcel Destombes. 1987. *The Earliest Printed Maps 1472-1500*. Edited by International Geographical Union Working Group on Early Maps. London: British Library.

Careri, Francesco. 2003. *Walkscapes: Walking as an Aesthetic Practice*. Barcelona: Aleu SA.

Carter, Paul. 1987. *The Road to Botany Bay: A Spatial History*. London: Faber and Faber.

Castells, Manuel. 1983. *The City and the Grassroots: A Cross-cultural Theory of Urban Social Movements*. London: Edward Arnold.

———. 1996. *The Rise of the Network Society. Vol. 1, The Information Age: Economy, Society and Culture*. Cambridge, MA: Blackwell.

Certeau, Michel de. 1984. *The Practice of Everyday Life*. Translated by Steven Rendell. Berkeley, California: University of California Press.

Chambers, Iain. 1994. *Migrancy, Culture, Identity*. London, UK: Routledge.

Chang, T C. 2005. "Place, Memory and Identity: Imagining 'New Asia'." *Asia Pacific Viewpoint* no. 46 (3):247-253.

Chattopadhyay, Swati. *A Critical History of Architecture in a Post-Colonial World: A View from Indian History* (v6n1.05) [internet]. Architronic 1997 [cited 27/09/2004. Available from http://architronic.saed.kent.edu/v6n1/v6n1.05a.html.

Chen, Kuo-wei. 2002. Meaning in Architectural and Urban Space of the Penang Kongsi Enclave. Paper read at The Penang Story - International Conference 2002, at Penang Malaysia.

Chihara, Daigoro. 1996. *Hindu-Buddhist Architecture in Southeast Asia*. Leiden: E. J. Brill.

Choay, Françoise. 1980. *The Rule and The Model*. Edited by Denise Bratton. English Translation ed. USA: The Massachussetts Institute of Technology. Original edition, Editions du Seuil. Reprint, 1997.

Christensen, C C. 1970. "Luther's theology and the uses of religious art." *The Lutheran Quarterly* no. 22:147-165.

Churchill, Robert R. 2004. "Urban Cartography and the Mapping of Chicago." *The Geographical Review* no. 94 (1):1-22.

Clammer, John. 2002. "Globalisation, Class, Consumption and Civil Society in South-east Asian Cities." *Urban Studies* no. 40 (2):403-419.

Clarke, R E. 1976. *Land and Neighbourhood as Features of Malaya Urbanism*, University of British Columbia Vancouver, Canada.

Cleaver, K C. 1989. "Adam Smith on Astronomy." *History of Science* no. 27:211-218.

Clifford, James. 1999. *Routes: Travel and Translation in the Late Twentieth Century*. Cambridge, Massachusetts: Harvard University Press.

Clulee, N. H. . 1988. *John Dee's Natural Philosophy: Between Science and Religion*. London and New York: Routledge.

Coedes, Georges. 1968. *The Indianised States of Southeast Asia*. Honolulu.

Cohn, Bernard S 1988. The Anthropology of a Colonial State and Its Forms of Knowledge. Paper read at Wenner-Gren conference on Tensions of Empire: Colonial Control and Visions of Rule, Spain.

Coleman, Nathaniel 2005. *Utopias and architecture* New York: Routledge.

Collins, George R. , and Christiane Crasemann Collins. 1965. *Camillo Sitte and the birth of modern city planning* London: Phaidon.

Colquhoun, Alan. 1997. "The Concept of Regionalism." In *Postcolonial Space(S)*, edited by Gülsüm Baydar Nalbantoğlu and Chong Thai Wong, 13-24. New York: Princeton Architectural Press.

Comito, T. 1971. "Renaissance Gardens and the Discovery of Paradise." *Journal of the History of Ideas* no. 32:483-506.

Contreras, Rafael. 2010. "Crackology." *Kerb 18* (PlastiCity FantastiCity: International Design Competition):22-29.

Cook, Andrew S. 1985. "Alexander Dalrymple and John Arnold: Chronometers and the Representation of Longitude on East India Company Charts." *Vistas in Astronomy* no. 28:189-95.

———. 1985. More By Accident Than Design: The Development of Topographical Mapping in India in the Nineteenth Century. In *Eleventh International Conference on the History of Cartography*. Ottowa, Canada.

Coomaraswamy, A. K. 1993. *Yaksas: Essays in the water cosmology*. Edited by Paul Schroeder. Delhi: Oxford University Press.

Corbusier, Le. 1929. *The City of Tomorrow and Its Planning*. Translated by Fredrich Etchells from *Urbanisme* 8th Edition. London: John Rodher. Original edition, 1929. Reprint, 1947 by Architectural Press.

———. 1948. *Concerning Town Planning*. Translated by Clive Entwhistle from *Propos d'urbanisme*. London: Architectural Press.

———. 1973. "The Athens Charter." In. New York: Grossman Publishers.

Cordua, Christian Hermansen. 2010a. *Manifestoes and Transformations in the Early Modernist City*. Surrey, England: Ashgate.

———. 2010b. "Prologue." In *Manifestoes and Transformations in the Early Modernist City*, edited by Christian Hermansen Cordua, 1-5. Surrey, GBR: Ashgate Publishing Group.

Cornell, T J. 1995. *The Beginnings of Rome*. London Routledge.

Corner, James. 1992. "Representation and Landscape." *Word & Image* no. VIII (3):243-75.

———. 1999a. "The Agency of Mapping: Speculation, Critique and Invention." In *Mappings*, edited by Denis Cosgrove, 213-252. London: Reaktion Books Ltd.

———. 1999b. *Recovering Landscape: Essays in Contemporary Landscape Theory*. USA: Princeton Architectural Press.

———. 2006. "Terra Fluxus." In *The Landscape Urbanism Reader*, edited by Charles Waldheim, 023-033. New York: Princeton University Press.

———. 2010. "Landscape Urbanism in the Field." *Topos: European Landscape Magazine* no. 71 (Landscape Urbanism):25-29.

Cosgrove, Denis. 1985. Geography and the mathematical arts: considerations on humanism, the occult and geography in the late Renaissance. Paper read at Annual Conference of the Institute of British Geographers, at University of Leeds.

———. 1992. "Mapping New Worlds: Culture and Cartography in Sixteenth-Century Venice." *Imago Mundi* no. 44:65-89.

———. 1999a. *Mappings*. London: Reaktion Books Pty Ltd.

———. 1999b. "Introduction: Mapping Meaning." In *Mappings*, edited by Denis Cosgrove, 1-23. London: Reaktion Books Ltd.

———.2000. Extra-Terrestrial Geography: Cosmography Before and After Von Humboldt. In *T. A. V. H. L. Series, The Alexander Von Humboldt Lecture*, edited by UCLA Department of Geography. California, LA: UCLA.

———.2003. "Globalism and Tolerance in Early Modern Geography." *Annals of the Association of American Geographers* no. 93 (4):852-870.

———. 2010. *Geography and Vision: Seeing, Imagining and Representing the World*. London, New York: I. B Tauris.

Crary, Jonathen. 1990. *Techniques of the Observer: On Mission and Modernity in the Nineteenth Century*. Cambridge, MA: The MIT Press.

Cresswell, Tim. 1997. "Imagining the Nomad: Mobility and the Postmodern Primitive." In *Space and Social Theory: Interpreting Modernity and Postmodernity*, edited by G. Benko and U. Strohmayer. Oxford: Blackwell Publishers Inc.

Cresswell, Tim, and Ginette Verstraete. 2003. *Mobilizing Place, Placing Mobility: The Politics of Representation in a Globalized World (Thamyris/Intersecting 9)*: Rodopi.

Daskalakis, George, Charles Waldheim, and Jason Young. 2001. *Stalking Detroit*. Barcelona: ACTAR.

Day, Tony. 2002. *Fluid Iron: State Formation in Southeast Asia*. Honolulu.

Day, Tony, and Craig J Reynolds. 2000. "Cosmologies, Truth Regimes, and the State in Southeast Asia." *Modern Asian Studies* no. 34 (1):1-55.

De Bernardi, Jean Elizabeth. 2004. *Rites of Belonging: Memory, Modernity and Identity in a Malaysian Chinese Community*. Stanford, California: Stanford University Press.

de Certeau, Michel. 1984. *The Practice of Everyday Life*. Translated by Steve Rendall. Berkeley: University of California Press.

Dear, Michael J. 2000. *The Postmodern Urban Condition*. Oxford: Blackwell.

Dear, Michael J., and Steven Flusty. 2001. *The Spaces of Postmodernity: Readings in Human Geography*. Malden, Massachussetts: Blackwell Publishers.

Deckker, Thomas. 2000. "The Modern City Revisited [electronic resource]." In. New York: Spon Press [Imprint] Routledge Florence : Taylor & Francis Group [Distributor]. http://library.adelaide.edu.au/item/1521012.

Delano Smith, Catherine. 1990. "Maps as Art and Science: Maps in Sixteenth Century Bibles." *Imago Mundi* no. 42:65-83.

———. 1994. "Prehistoric Cartography in Asia." In *The History of Cartography: Cartography in the Traditional East and Southeast Asian Societies*, edited by J. B. Harley and David Woodward, 1-22. Chicago, USA: The University of Chicago Press Pty Ltd.

Delano-Smith, Catherine, and Elizabeth Morley Ingram. 1991. *Maps in Bibles 1500-1600: An Illustrated Catalogue*. Geneva: Librairie Droz.

Deleuze, Gilles, and Felix Guattari. 1986. *Nomadology: The War Machine*. Translated by Brian Massumi. New York: University of Minnesota Press. Original edition, Mille Plateaux, Paris, Editions de Minuit.

———. 1987. *A Thousand Plateaus: Capitalism and Schizophrenia*. Translated by Brian Massumi. Minneapolis: University of Minnesota Press.

Desai, W. S. 1936. "A Map of Burma (1795) by a Burmese Slave." *Journal of Burma Research Society* no. 26 (3):147-52.

Dilke, O. A. W. 1998. *Greek and Roman Maps*. London: Johns Hopkins University Press.

Dirks, Nicholas. 1993. "Colonial Histories and Native Informants: Biography of an Archive." In *Orientalism and the Postcolonial Predicament: Perspective on South Asia*, edited by Carol Breckenridge and Peter Van der Veer. Philadelphia: University of Pennsylvania Press.

Dodds, Jerrilynn D. "The Great Mosque of Cordoba." In *Al-Andalus*, edited by J Dodds. New York: Metropolitan Museum of Art.

Dorling, Daniel, and David Fairbairn. 1997. *Mapping: Ways of Representing the World*: Prentice Hall.

Dostoğlu, Neslihan Turkun 1986. *Architectural deterministic thinking in the development of urban utopias, 1848-1947*, Architecture, University of Pennsylvania, Ann Arbor, Mich.

Dovey, Kim. 1999. *Framing Places: Mediating Power in Built Form*. Edited by T. A. Markus and A. D. King, *Architext Series*. London: Routledge.

Duncan, James. 1972. *Milton's Earthly Paradise: A Historical Study of Eden*. Minneapolis: University of Minnesota Press.

Eade, J. C. . 1989. *Southeast Asian Ephemeries: Solar and Planetary Positions, A. D. 638-2000*. Ithaca, New York: Cornell Southeast Asia Program.

Eaton, Ruth. 2002. *Ideal Cities: Utopianism and the (Un) Built Environment*. London: Thames and Hudson.

Edney, Matthew. 1994. "Mathematical Cosmography and the Social Ideology of British Cartography 1780-1820." *Imago Mundi* no. 46:101-16.

———. 1997. *Mapping an Empire: The Geographical Construction of British India 1765-1843*. Chicago: University of Chicago Press.

Edney, Matthew H. 1999. "Reconsidering Enlightenment Geography and Map Making: Reconnaissance, Mapping, Archive." In *Geography and Enlightenment*, edited by Charles W J Withers and David N Livingstone, 165-198. Chicago USA: The University of Chicago Press.

Ehrenberg, Ralph E. 2006. *Mapping the World: An Illustrated History of Cartography*. Washington D.C.: National Geographic Society.

El-Khoury, R., and Edward Robbins. 2003. *Shaping the City: Studies in History, Theory and Urban Design*. New York: Routledge.

Elliott, James. 1987. *The City in Maps: urban mapping to 1900*. London: The British Library Board.

Emmerson, Donald K. 1984. "Southeast Asia: What's in a Name?" *Journal of Southeast Asian Studies* no. 15:1-21.

Evenson, N. 1966. *Chandigarh*. Berkeley and Los Angeles: University of California Press.

Evers, Hans-Dieter. 1977. "The Culture of Malaysian Urbanisation: Malay and Chinese Conceptions of Space." *Urban Anthropology* no. 6 (3):205-16.

Evers, Hans-Dieter, and Rudiger Korff. 2000. *Southeast Asian Urbanism: The Meaning and Power of Social Space*. Germany/USA: Lit Verlag/St Martin's Press.

Fall, Juliet. 2005. *Drawing the Line: Nature, Hybridity and Politics in Transboundary Spaces*. Edited by Doris Wastl-Walter, *Border Region Series*. Aldershot, England: Ashgate Publishing Limited.

Fell, R T. 1988. *Early Maps of South-East Asia*. Singapore: Oxford University Press Pty Ltd.

Fisher, Charles A. 1964. *South-east Asia: A Social, Economic and Political Geography*. London: Methuen.

Fishman, Robert. 1977. *Urban Utopias in the Twentieth Century: Ebenezer Howard, Frank Lloyd Wright, and Le Corbusier*. New York: Basic Books Inc. Publishers.

Fletcher, D H. 1995. *The Emergence of Estate Maps. Christ Church, Oxford c. 1600-1840*. Oxford: Oxford University Press.

Forbes, Dean. 1996. *Asian Metropolis: Urbanisation and the Southeast Asian City*. Melbourne, Australia: Oxford University Press.

Forbes, Eric G. 1974. *The Birth of Scientific Navigation: The Solving in the Eighteenth Century of the Problem of Finding Longitude at Sea*. Greenwich: National Maritime Museum.

Forman, Richard, T. T. 2009. "Urban Ecology and the Arrangement of Nature in Urban Regions." In *Ecological Urbanism* edited by Mohsen Mostafavi and Gareth Doherty, 312-323. Harvard University Graduate School of Design USA: Lars Muller Publishers.

Foucault, Michel. 1970. *The Order of Things: An Archaeology of the Human Sciences*. New York: Random House.

Fournier, Colin. 2001. "Webbed Babylon." *Architectural Design* no. 71 (3).

Fowle, Frances, and Richard Thomson. 2003. *Soil and Stone: Impressionism, Urbanism, Environment*: Aldershot Ashgate.

Fraser, Valerie. 2000. "Cannibalizing Le Corbusier: The MES Gardens of Roberto Burle Marx." *Journal of the Society of Architectural Historians* no. 59 (2):180-193.

Freart, Roland sieur de Chambray. 1707. *A parallel of the ancient architecture with the modern : in a collection of ten principal authors who have written upon the five orders* Second ed. London D. Brown, J. Walthoe, B. Took and D. Midwinter.

Fu, Shen C. Y. , and et al. 1977. *Traces of the Brush: Studies in Chinese Calligraphy*. New Haven: Yale University Press.

Fung, Stanislaus. 2005. "The Language of Cultural Memory in Chinese Gardens." In *Structure and Meaning in Human Settlements*, edited by Tony Atkin and Joseph Rykwert, 123-134. Philadelphia: University Pennsylvania Museum of Archaeology and Anthropology.

Furnivall, J S. 1980. "Plural Societies." In *Sociology of Southeast Asia: Readings on Social Change and Development*, edited by Hans-Dieter Evers, 86-96. Kuala Lumpur: Oxford University Press.

Gandelsonas, Mario. 1991. *Urban Text.* Cambridge, Massachussetts: Chicago Institute for Architecture & Urbanism (CIAU) MIT Press.

Gandelsonas, Mario, Ackbar Abbas, and Christine Boyer. 2002. *Shanghai Reflections: Architecture, Urbanism and the Search for an Alternative Modernity*. New York: Princeton Architectural Press.

Giddens, Anthony. 1990. *The Consequences of Modernity*. Cambridge MA: Polity.

Glacken, C. 1967. *Traces on the Rhodian shore. Nature and culture in Western thought from ancient times to the end of the eighteenth century*. Berkeley: University of California Press.

Goetzmann, William H. 1986. *New Lands, New Men: America and the Second Great Age of Discovery*. New York: Viking.

Goh, Beng-Lan. 2002. *Modern Dreams: An Inquiry into Power, Cultural Production and the Cityscape in Contemporary Penang Malaysia*. New York: Ithaca, South-East Asia Programme, Cornell University.

Goh, Ban Lee. 2004. *Urban Planning in Malaysia: History, Assumptions and Issues*. Malaysia: Delta Editions.

Goh, Robbie B. H., and Shawn Wong. 2004. *Asian Diasporas: Cultures, Identities, Representations*. Hong Kong: Hong Kong University Press.

Goh, Robbie B. H., and Brenda S. A. Yeoh. 2003. *Theorising the Southeast Asian City as Text: Urban Landscapes, Cultural Documents, And Interpretative Experiences*. Singapore: World Scientific Publishing Co. Inc.

Gordon, B. L. 1971. "Sacred Directions, Orientation, and the Top of the Map." *History of Religions* no. 10 (3):211-227.

Gosling, L. A. Peter. 1977. "Contemporary Malay Traders in the Gulf of Thailand." In *Economic Exchange and Social Interaction in Southeast Asia*, edited by K L Hutterer, 73-95. Ann Arbor, USA: Centre for South and Southeast Asian Studies, University of Michigan.

Grabar, Oleg. 1969. "The Architecture of Middle Eastern Cities: The Case of the Mosque." In *Middle Eastern Cities in Comparitive Analysis*, edited by Ira Lapidus. USA: University of California Press.

———. 1973. *The Formation of Islamic Art*. USA: New Haven.

———. 1998. *The Shape of the Holy*. USA: Harvard University Press.

Graham, Stephen, and Simon Marvin. 2001. *Splintering Urbanism: Networked Infrastructures, Technological Mobilities and the Urban Condition*. London: Routledge.

Gregory, Derek. 1994. *Geographical Imaginations*. Cambridge MA & Oxford UK: Blackwell Publishers.

Griffith, Robert, and Carol Thomas. 1981. *The City-State in Five Cultures*. Santa Barbara: ABC-Clio.

Groslier, Bernard Philippe. 1962. *Indochina: Art in the melting-pot of races*. London: Methuen.

Guillen, Mauro F. 2006. *The Taylorised Beauty of the Mechanical: Scientific Management and the Rise of Modernist Architecture*. Princeton and Oxford: Princeton University Press.

Gunn, Geoffrey C. 2003. *First Globalisation - The Eurasian Exchange, 1500-1800*. Edited by Mark Selden, *World Social Change*. Maryland, USA: Rowman & Littlefield Publishers INC.

Hahn, Roger. 1989. "New Considerations on the Physical Sciences of the Enlightenment Era." *Studies on Voltaire and the Eighteenth Century* no. 264:789-796.

Hakim, Besam. 1978. Sidi Bou Said, Tunisia. Halifax, Nova Scotia, Canada: Nova Scotia Technical College, Canada.

Hakim, Besim Selim. 1986. *Arabic-Islamic Cities: Building and Planning Principles*. 1st ed. London: Kegan Paul International Ltd. Original edition, 1979. Reprint, 1988.

Hall, Daniel George Edward 1968. *A History of South-east Asia*. 3rd ed. New York: St Martin's Press.

Hall, Kenneth. 1985. *Maritime Trade and State Development in Early Southeast Asia*. Honolulu: University of Hawaii Press.

Hall, Kenneth, and John Whitmore. 1976. *Explorations in Early Southeast Asian History: The Origins of Southeast Asian Statecraft* Ann Arbor.

Hall, Peter Geoffrey. 1996. *Cities of Tomorrow: An Intellectual History of Urban Planning and Design in the Twentieth Century*. updated edition ed.

Oxford: Blackwell Publishers. Original edition, 1988. Reprint, 1997,1998,1999,2000,2001.

Hall, Stephen. 1992. *Mapping the Next Millenium: The Discovery of New Geographies*. New York: Random House.

Halton, Eugene 1995. "The Modern Error: Or, the Unbearable Enlightenment of Being." In *Global modernities* edited by Mike Featherstone, Scott Lash and Roland Robertson. London ; Thousand Oaks, Calif. : Sage Publications.

Hamnett, C. 1994. "Social polarisation in global cities." *Urban Studies* no. 31 (3):401-24.

Haneda, Masashi, and Toru Miura. 1994. *Islamic Urban Studies: Historical Review and Perspectives*. 1st ed. London, New York: Kegan Paul International.

Hanisch, Ruth. 2010. "Camillo Sitte: City Planning According to Artistic Principles, Vienna 1889." In *Manifestoes and Transformations in the Early Modernist City*, edited by Christian Hermansen Cordua, 125-136. Surrey, England: Ashgate.

Hannigan 1995. "Postmodern Architecture and Urban Design." *Current Sociology* no. 43:165-172. doi: 10.1177/001139295043001013.

Harley, J. B. 1983. "Meaning and ambiguity in Tudor cartography." In *English Map-making 1500-1650*, edited by S Tyacke, 22-45. London: The British Library.

———. 1987 "Maps, Knowledge and Power." In *The Iconography of Landscape: Essays on the Symbolic Representation, Design and Use of Past Environments*, edited by Denis Cosgrove and Stephen J Daniels, 277-312. Cambridge: Cambridge University Press.

———. 1992. "Deconstructing the Map." In *Writing Worlds*, edited by Trevor J Barnes and James Duncan, 231-47. London: Routledge.

Harley, J. B., and David Woodward. 1992. *The History of Cartography: Cartography in the Traditional Islamic and South Asian Societies*. Edited by Joseph E Schwartzberg, Gerald R Tibbetts and Ahmet T Karamustafa. 6 vols. Vol. 2, Book 2. Chicago: University of Chicago Press Pty Ltd.

Harley, J B, and David Woodward. 1994. *The History of Cartography: Cartography in the Traditional East and Southeast Asia*. 2 vols. Vol. 2,

Book 2, *The History of Cartography*. Chicago, USA: The University of Chicago Press Pty. Ltd.

Harris, Steven, and Deborah Berke. 1997. *Architecture of the Everyday*. New York: Princeton Architectural Press.

Harris, W. V. 1971. *Rome in Etruria and Umbria*. Oxford: Oxford University Press.

Harvey, David. 1969. *Explanation in geography* London: Edward Arnold.

———. 1989. *The Condition of Postmodernity*. first ed. Oxford: Basil Blackwell Ltd.

———. 2000. *Spaces of Hope*. Edinburgh: Edinburgh University Press.

Harvey, P. D. A. 1980. *The History of Topographical Maps: Symbols, Pictures and Surveys*. London: Thames and Hudson.

———. 1994. *Maps in Tudor England*. London: University of Chicago Press.

Heine-Geldern, Robert. 1956. *Conceptions of State and Kingship in Southeast Asia*. Ithaca: Cornell University Southeast Asia Program.

Held, David, Anthony McGrew, D Goldblatt, and J Perraton. 1999. *Global Transformations: Politics, economies and culture*. Cambridge MA: Polity.

Hendersen, John B. 1994. "Chinese Cosmographical Thought: The High Intellectual Tradition." In *The History of Cartography: Cartography in the Traditional East and Southeast Asian Societies*, edited by J B Harley and David Woodward, 202-227. Chicago: The University of Chicago Press Ltd.

Herscher, Andrew. 2003. "Städtebau as Imperial Culture: Camillo Sitte's Urban Plan for Ljubljana." *The Journal of the Society of Architectural Historians* no. 62 (2):212-227.

Hewitt, Mark. 1985. "Representational Forms and Modes of Conception: An Approach to the History of Architectural Drawing." *Journal of Architectural Education* no. 39 (2):2-9.

Heynen, Hilde. 1999. *Architecture and Modernity: A Critique*. Cambridge, Massachussetts: MIT Press.

Hilberseimer, L. 1955. *The Nature of Cities*. Chicago, USA: Paul Theobald and Co.

Hindman, S. 1977. *Text and image in fifteenth century illustrated bibles*. Leiden: E. J. Brill.

Hindman, Sandra. 1977. *Text and Image in the Fifteenth Century*. Leiden: E. J. Brill.

Holt-Jensen, A. 1980. *Geography: It's History and Concepts*. London: Harper and Row.

Home, Robert. 1997. *Of Planting and Planning: The Making of British Colonial Cities*. London: Spon.

Hosagrahar, Jyoti. 2002. "South Asia: Looking Back, Moving Ahead - History and Modernisation." *Journal of Society of Architectural Historians* no. 61 (3):355-369.

Howard, Ebenezer. 1898. *To-Morrow: A Peaceful Path to Real Reform*. 2nd ed, *Garden Cities of To-morrow*. London: Swan Sonnenschein.

Howse, Derek, and Michael Sanderson. 1973. *Sea Chart*. UK: David & Charles PLC.

Hsu, Mei-Ling. 1978. "The Han Maps and Early Chinese Cartography." *Annals of the Association of American Geographers* no. 68 (1):9-17.

Hughes, Johnathon, and Simon Sadler. 2000. *Non-Plan: Essays on Freedom, Participation and Change in Modern Architecture and Urbanism*. London: Oxford University Press.

J, Coulton J. 1977. *Ancient Greek Architects At Work*. Ithaca, NY: Cornell University Press.

Jackson, Richard Joseph 2004. *Urban Sprawl and Public Health*. USA: Island Press.

Jacob, Christian. 1999. "Mapping in the Mind: The Earth from Ancient Alexandria." In *Mappings*, edited by Denis Cosgrove, 24-49. London: Reaktion Books Ltd.

Jacoby, Russell. 1999. *The End of Utopia: Politics and Culture in an Age of Apathy*. New York: Basic Books.

Jenkins, Eric J. 2008. *To Scale: One Hundred Urban Forms*. NY, USA: Routledge.

Jormakka, Kari 2006. *The art of the city : from Camillo Sitte to today* Tampere: Tampere University of Technology.

Jumsai, Sumet. 1988. *Naga: Cultural Origins of Siam and the West Pacific*. Singapore: Oxford University Press.

Junker, Laura Lee. 2006. "Population Dynamics and Urbanism in Premodern Island Southeast Asia." In *Urbanism in the Preindustrial World: Cross-Cultural Approaches*, edited by Glenn R Storey, 203-232. Tuscaloosa,USA: The University of Alabama Press.

Kahn, C H. 1960. *Anaximander and the Origins of the Greek Cosmology*. New York: Hackett.

Kalpagam, U. 1995. "Cartography in Colonial India." *Economic and Political Weekly* no. 30 (30):PE87-PE98.

Karamustafa, Ahmet T. 1992. "Introduction to Islamic Maps." In *History of Cartography Volume 2, Book 1: Cartography in the Traditional Islamic and South Asian Societies*, edited by J B Harley and David Woodward, 3-11. Chicago: The University of Chicago Press.

Kelley, K B. 1976. "Dendritic Central-Place Systems and the Regional Organisation of Navajo Trading Posts." In *Regional Analysis: Vol. I Economic Systems*, edited by C Smith, 219-254. New York: Academic Press.

Kenneth, Powell. 2000. *City Transformed: Urban Architecture at the Beginning of the 21st Century*. London: Laurence King Publishing.

Khoo, Su Nin. 1993. *Streets of George Town Penang*. Kuala Lumpur, Malaysia: Janus Print and Resources.

Kien, Lai Chee. 2010. "Southeast Asian Spatial Histories and Historiographies." *The Journal of the Society of Architectural Historians, Australia and New Zealand* no. 19 (2):82-105.

King, Anthony. 1990. *Global Cities: Post-Imperialism and the Internationalisation of London*. London UK: Routledge.

———. 1976. *Colonial Urban Development, Culture, Social Power and Environment*. London: Routledge and Kegan Paul.

———. 1982. "Colonial Architecture and Urban Development: Reconversion of Colonial Typologies in Independent India." *Lotus International* no. 34:47-59.

———. 2004. *Spaces of Global Cultures: architecture, urbanism, identity., Architext Series*. London UK: Routledge.

King, David A. 1999. *World Maps For Finding The Direction and Distance to Mecca: Innovation and Tradition in Islamic Science*. Edited by H Daiber and D Pingree. Vol. xxxvi, *Islamic Philosophy Theology and Science*. Leiden, The Netherlands: Brill.

Konvitz, Josef W. 1987. *Cartography in France, 1660-1848: Science, Engineering and Statecraft* Chicago: The University of Chicago Press.

Koolhaas, Rem. 1978. *Delirious New York* London Thames and Hudson.

———. 2004. *Content*. Cologne: Taschen.

———. 2010. "Advancement versus Apocalypse." In *Ecological Urbanism*, edited by Mohsen Mostafavi and Gareth Doherty, 56-71. Baden, Switzerland: Lars Muller Publishers & the President and Felllows of Harvard College.

Kostof, Spiro. 1991. *The City Shaped: Urban Patterns and Meanings Through History*. London: Thames and Hudson.

———. 1992. *The City Assembled: The Elements of Urban Form Through History*. London: Thames and Hudson.

Kramrisch, Stella. 1976. *The Hindu Temple*. 2nd Edition ed. 2 vols. Vol. 1. Delhi: Motilal Banarsidass. Original edition, 1946 (Calcutta: University of Calcutta). Reprint, 1976.

Kratoska, Paul H., Remco Raben, and Henk Schulte Nordholt. 2005. *Locating Southeast Asia: Geographies of Knowledge and Politics of space*. Singapore: Singapore University Press, National University of Singapore.

Krause, Linda. 2003. *Global Cities: Cinema, Architecture and Urbanism in a Digital Age, New Directions in International Studies*. New Brunswick, New Jersey: Rutgers University Press.

Kroll, Lucien. 1988. *Buildings and Projects*. London: Thames and Hudson Inc.

Kunstler, J. Howard. 2005. *The long emergency*. Canada: Atlantic Monthly Press.

Kusmiadi, Rachmat. 1977. A Brief History of Cartography in Indonesia. Paper read at Seventh International Conference on the History of Cartography, 7-11 August 1977, at Washington D. C.

Kusno, Abidin. 2000. *Behind the Postcolonial: Architecture, urban space and political cultures in Indonesia*. Edited by Thomas A. Markus and Anthony D. King, *Architext Series*. London: Routledge.

———. 2000. "Imagining Regionalism, Re-Fashioning Orientalism: Some Current Architectural Discourses in Southeast Asia." *Journal of Southeast Asian Architecture* no. 4:45-61.

———. 2010. "Tropics of discourse: notes on the re-invention of architectural regionalism in Southeast Asia in the 1980s." *Fabrications* no. 19 (2):58-73.

Lach, Donald F., and Edwin J. van Kley. 1993. *Asia in the Making of Europe, Volume III. A Century of Advance. Book Three Southeast Asia.* Vol. III. Chicago and London: The University of Chicago Press.

Lampl, P. 1968. *Cities and Planning in the Ancient Near East.* USA: New York.

Lapidus, Ira. 1987. *Muslim Cities in the Later Middle Ages*. USA: Berkeley.

Lassner, L. "The Caliph's Personal Domain: The City Plan of Baghdad Re-Examined." In *The Islamic City*, edited by Hourani & Stern.

Lawrence, Roderick J. 1996. "The Multi-dimensional Nature of Boundaries: An Integrative Historical Perspective." In *Setting Boundaries*, edited by Deborah Pellow. Westport, USA: Bergin & Garvey.

Laxton, Paul. 1976. "The Geodetic and Topographical Evaluation of English County Maps, 1740-1840." *The Cartographic Journal* no. 13:37-54.

———. 2001. *J. B. Harley: The New Nature of Maps: Essays in History of Cartography.* USA: John Hopkins University Press.

Leblanc, Linda, and Jacques Coulon. 1993. *Paysages*. France: Publications du Montieur.

Leccese, Michael, and Kathleen McCormick. 2000. *Charter of the New Urbanism.* New York: McGraw Hill.

Ledyard, Gari. 1994. "Cartography in Korea." In *The History of Cartography: Cartography in the Traditional East and Southeast Asian Societies*, edited by J. B Harley and David Woodward, 235-345. Chicago: University of Chicago Press.

Leed, Erich J. 1991. *The Mind of the Traveller: From Gilgamesh to Global Tourism.* New York: Basic Books.

Lefaivre, Liane, and Alexander Tzonis. 2003. *Critical Regionalism: Architecture and Identity in a Globalised World.*

Lefebrve, Henri. 1991. *The Production of Space*. Oxford, UK.: Blackwell.

———. 1994. *Everyday Life in the Modern World.* Translated by Sacha Rabinovitch. 3rd ed. New Jersey, USA: Transaction Publishers. Original edition, 1984 with this material, original material 1971.

———. 2003. *The Urban Revolution*. Translated by Robert Bononno. Minneapolis: University of Minnesota.

Lefebrve, Henri, Stuart Elden, and Gerald Moore. 2004. *Rythmnanalysis: Space, time and Everyday Life*. New York: Continuum.

Ley, David. 1989. "Modernism, postmodernism and the struggle for place." In *The Power of Place: Bringing Together the Geographical and Sociological Imaginations*, edited by John Agnew and James Duncan, 44-65. London: Unwin Hyman.

Ley, David, and Caroline Mills. 1993. "Can there be a postmodernism of resistance in the urban landscape?" In *The Restless Urban Landscape*, edited by Paul Knox, 255-78. Englewood Cliffs, NJ: Prentice Hall.

Lieberman, Victor. 2003. *Strange Parallels: Southeast Asia in a Global Context, 800-1830*. 2 vols. Vol. 1: Integration on the Mainland. Cambridge.

Lim, William Siew Wai. 1998. *Asian New Urbanism and Other Papers*. Singapore: Select Books.

———. 2003. *Alternative (Post)modernity: An Asian Perspective*. Singapore: Select Publishers.

Lim, William Siew Wai, and Philip Motha. 1980. *An Alternative Urban Strategy*. Singapore: DP Architects.

Lim. William, S. W., and Jiat-Hwee Chang. 2012. *Non West Modernist Past: On Architecture and Modernities*. Singapore: World Scientific Publishing Co. Pte. Ltd.

Livingstone, David N. 1988. "Science, Magic, and Religion: A Contextual Reassessment of Geography in the Sixteenth and Seventeenth Centuries." *History of Science* no. 26:269-94.

———. 1990. "Geography, Tradition and the Scientific Revolution: An Interpretative Essay." *Transactions of the Institutes of British Geographers. New Series* no. 15 (3):359-373.

———. 1992. *The preadamite theory and the marriage of science and religion*. Philadelphia: American Philosophical Society.

Low, Setha M. 1999. *Theorising the City: The New Urban Anthropology Reader*. New Brunswick, N.J.: Rutgers University Press.

———. 2000. *On the Plaza: The Politics of Public Space and Culture*. Austin: University of Texas Press.

Low, Setha M., and Denise Lawerence-Zauaniga. 2003. *The Anthropology of Space and Place: Locating Culture*. Malden, Massachussetts: Blackwell Publishers.

Lowenthal, David, and Martyn J Bowden. 1976. *Geographies of the Mind: Essays in Historical Geosophy*. New York: Oxford University Press.

Luckerman, F, and P Porter. "The Geography of Utopia." In *Geographies of the Mind: Essays in Historical Geography* edited by David Lowenthal and Martin Bowden.

MacDonald, W. 1986. *The Architecture of the Roman Empire, Volume II An Urban Appraisal*. USA: Yale University Press.

MacIntyre, Alasdair. 1990. *Three Rival Versions of Moral Enquiry: Encyclopedia, Geneaolgy, and Tradition*. Notre Dame IN: Notre Dame University Press.

Madanipour, Ali 2007. *Designing the city of reason : foundations and frameworks* London: Routledge.

Mangani, G. 1998. "Abraham Ortelius and the hermetic meaning of the cordiform projection." *Imago Mundi* no. 50:59-83.

Mannikka, Eleanor. 1996. *Angkor Wat: Time, Space and Kingship*. Hawaii: University of Hawaii Press USA.

March, Benjamin. 1931. "Linear Perspective in Chinese Painting." *Eastern Art* no. 3:113-39.

Marr, David, and Anthony Milner. 1986. *Southeast Asia in the 9th to 14th Centuries*. Singapore: Institute of Southeast Asian Studies.

Marshall, Richard. 2001. *Waterfronts in Post-Industrial Cities*. London: Spon Press.

———. 2003. *Emerging Urbanity: Global Projects in the Asia Pacific Rim*. London: Spon Press.

McClellan III, James E. 1992. *Colonialism and Science: Saint Domingue in the Old Regime* Baltimore and London: The Johns Hopkins Univ. Press.

McClellan, James E. III , and Francois Regourd. 2000. "The Colonial Machine: French Science and Colonization in the Ancien Regime." *Osiris, Nature and Empire: Science and the Colonial Enterprise* no. Vol. 15 (2nd Series):31-50.

McDonough, Tom 2009. *The situationists and the city*. London Verso.

McDonough, William, and Braungart Michael. 2002. *Cradle to Cradle – Remaking the way we make things* USA: North Point Press.

McGee, Terry. 1967. *The Southeast Asian City: A Social Geography of the Primate Cities of Southeast Asia* London: G Bell.

McGee, Terry, and Ira Robinson. 1995. *The Mega-Urban Regions of Southeast Asia*. Vancouver, Canada.

McGregor, Adrian. 2010. "Biocity: Emergent Sustainability." *Topos: European Landscape Magazine* no. 70 (Sustainability):70-75.

Meister, Michael W. 2000. "Mountains and Cities in Cambodia: Temple Architecture and Divine Vision." *International Journal of Hindu Studies* no. 4 (3):261-268.

Mendyk, S. 1989. *Speculum Britanniae: Regional Study, Antiquarianism and Science in Britain to 1700*. Toronto: Univeristy of Toronto.

Meyer, Jeffrey F. 1978. "Feng-Shui of the Chinese City." *History of Religions* no. 18 (November 1978):138-55.

Milgram, Stanley. 1970. "The Experience of Living in Cities: A psychological analysis." *Science* no. 167:1461-1468.

Mills, J. V. 1953. "Chinese Coastal Maps." *Imago Mundi* no. 11:151-168.

Molestina, J Pablo. 2004. "Greenurbia Or the Return of the Wolves." *Architectural Design* no. 74 (4):60-65.

Monmonier, Mark. 1991. *How to Lie with Maps*. Chicago: University of Chicago Press.

More, Thomas. 1524-1625 (2008). "Utopia." In. Adelaide: The University of Adelaide Library. Link to e-Resource: http://ebooks.adelaide.edu.au/m/more/thomas/m83u/.

Moron, Eleanor 1977. "Configurations of Time and Space at Angkor Wat." *Studies in Indo-Asian Art and Culture* no. 5:217-261.

Mostafavi, Mohsen. 2010. "Why Ecological Urbanism? Why Now?" *Topos: European Landscape Magazine* no. 71 (Landscape Urbanism):30-35.

Mostafavi, Mohsen, Ciro Najile, and Architectural Association. 2003. *Landscape Urbanism: A Manual for the Machinic Landscape*. London: Architectural Association.

Mumford, Lewis. 1967. "Utopia, the city and the machine." In *Utopias and Utopian Thought: A Timely Appraisal*, edited by Frank E Manuel, 3-24. Boston, MA: Beacon Press.

N, Goodman. 1968. *Languages of Art: An Approach to a Theory of Symbols*. New York: Bobbs-Merril.

Naddeo, Barbara Ann. 2004. "Topographies of Difference: Cartography of the City of Naples 1627-1775." *Imago Mundi* no. 56 (Part 1):23-47.

Naerssen, F. H. 1976. "Tribute to the God and Tribute to the King." In *Southeast Asian History and Historiography: Essays presented to D. G. E. Hall*, edited by C. D. Cowan and O W Wolters. Ithaca: Cornell University Press.

Naisbitt, John. 1996. *Megatrends Asia: The Eight Asian Megatrends that are Changing the World*. London: Nicholas Brealey Publishing.

Nalbantoğu, Gülsüm Baydar, and Thai Wong Chong. 1997. *Postcolonial Space(S)*. USA: Princeton Architectural Press.

Nasr, Seyyed Hosein. 1966. *Ideals and Realities of Islam*. London: George Allen and Unwin Ltd.

Nebenzahl, Kenneth. 1987. *Maps of the Bible Lands*. London: Times Books.

Needham, Joseph. 1954. *Science and Civilisation in China* Vol. 3. Cambridge: Cambridge University Press.

Newman, Peter , and Isabella Jennings. 2008. *Cities as sustainable ecosystems*. . USA: Island Press

Nichols, Julie. 2002. Discrepant Urbanities: A Reading of the Urban History of Penang. Paper read at Southern Crossings - Proceedings for the Sixth Australasian Urban History/Planning History Conference, February 2002, at Auckland, New Zealand.

———. 2006. "Nomadic Urbanities: Constant's New Babylon and the Contemporary City." *Graduate Journal of Asia-Pacific Studies* no. 4 (2):29-52.

———. 2007. Mapping the City: Revisiting the map in contemporary urban design. Paper read at Society of Architectural Historians for Australia and New Zealand (SAHANZ), September 2007, at Adelaide, Australia.

———.2010. Imaginative Geographies: Cosmography and Cartography in Pre-modern Southeast Asia. Paper read at Imagining 27th International

SAHANZ Conference published in the Proceedings, The University of Newcastle, Australia, 30 June-2 July.

———.2012. Le Corbusier and Chandigarh: Mapping Utopia and the Emergence of Modern Urbanism," published in the Proceedings of the Theoretical Currents II – Architecture and Its Geographical Horizons Conference, University of Lincoln, UK, 4-5 April.

Nicolet, Claude. 1991. Space, Geography, and Politics in the Early Roman Empire. In *Jerome Lectures*. Ann Arbor: University of Michigan Press.

Nordholt Schulze, Henk. 2011. "Indonesia in the 1950s: nation, modernity, and the post-colonial state." *Journal of the Humanities and Social Sciences of Southeast Asia and Oceania* no. 167 (4):386-400.

Noyes, John K. 1993. "The Representation of Spatial History." *Pretexts* no. 4 (2):120-27.

Nuti, Lucia. 1999. "Mapping Places: Chorography and Vision in the Renaissance." In *Mappings*, edited by Denis Cosgrove, 90-108. London: Reaktion Books.

O'Connor, Richard, A. 1983. *A Theory of Indigenous Southeast Asian Urbanism, Research Notes and Discussions Paper No. 38*. Singapore: Institute of Southeast Asian Studies.

Owen, Norman G. 2005. *The Emergence of Modern Southeast Asia*. Honolulu: University of Hawai'i Press.

P., Hardie. 1986. *Virgil's Aeneid: Cosmos and Imperium*. Oxford Clarendon Press, Oxford University Press.

Padron, Ricardo. 2002. "Mapping Plus Ultra: Cartography, Space and Hispanic Modernity." *Representations* no. 79 (Summer 2002):28-60.

Pagden, A. 1986. "The impact of the new world on the old: the history of an idea." *Renaissance and Modern Studies* no. 30:1-11.

Palmboom, Fritz. 2010. "Landscape Urbanism: Conflation or Coalition?" *Topos: European Landscape Magazine* no. 71 (Landscape Urbanism):43-49.

Pancaroğlu, Oya. 2003. "Signs in the Horizon: Concepts of Image and Boundary in a Medieval Persian Cosmography." *Anthropology and Aesthetics, Islamic Arts* no. 43 (Spring 2003):31-41.

Pandya, Vishvajit. 1990. "Movement and Space: Andamanese Cartography." *American Ethnologist* no. 17:775-97.

Pedley, Mary Sponberg. 1995. Commode, complet, uniforme, et suivi': Problems in Atlas Editing in Enlightenment France. Paper read at Editing in Early and Historical Atlases: Paapers Given at the Twenty-ninth Annual Conference on Editorial Problems, at University of Toronto, Canada.

Pellow, Deborah. 1996a. "Intimate Boundaries: A Chinese Puzzle." In *Setting Boundaries*, edited by Deborah Pellow. Westport USA: Bergin & Garvey.

———. 1996b. *Setting Boundaries: The Anthropology of Spatial and Social Organisation*. USA: Bergin & Garvey.

Peters, F E. 1986. *Jerasulem and Mecca: The Typology of the Holy City in the Near East*. USA: New York university Press.

Phillimore, Reginald Henry. 1956. "An Early Map of the Malay Peninsula." *Imago Mundi* no. 13:174-79.

Pickles, John. 1992. "Texts, Hermeneutics and Propaganda Maps." In *Writing Worlds*, edited by Trevor J Barnes and James Duncan. London: Routledge.

———. 2004. *A history of spaces: Cartographic reason, mapping and the geo-coded world*. London: Routledge.

Pinches, Michael. 1994. "Modernisation and the Quest for Modernity: Architectural Form, Squatter Settlements and The New Society in Manila." In *Cultural Identity and Urban Change in Southeast Asia: Interpretative Essays*, edited by Marc Askew and William S Logan, 13-42. Geelong, Victoria: Deakin University Press.

Pinder, David. 2001. "Utopian Transfiguration: The Other Spaces of New Babylon." *Architectural Design* no. 71 (3):15-19.

———. 2010. "Utopian Urbanism: Ideals, Practices and Prospects." In *Manifestoes and Transformations in the Early Modernist City* edited by Christian Hermansen Cordua. Farnham, Surrey GBR: Ashgate Publishing Group.

Pinto, John A. 1976. "Origins and Developments of the Ichnographic City Plan." *The Journal of the Society of Architectural Historians* no. 35 (no. 1):35-50.

Pizarro, Rafael E, Liang Wei, and Tridib Banerjee. 2003. "Agencies of Globalisation and Third World Urban Form: A Review." *Journal of Planning Literature* no. 18 (2):111-130.

Plaut, A. 1984. "Where is paradise? The Mapping of A Myth." *The Map Collector* no. 29:2-7.

Powell, Robert. 1999. *Rethinking the Skyscraper: The Complete Architecture of Ken Yeang*. London: Thames and Hudson.

Pratt, Mary Louise. 1992. *Imperial Eyes: Travel Writing and Transculturation.* London: Routledge.

Prest, John. 1981. *The Garden of Eden: the Botanic Garden and the Re-creation of Paradise*. New Haven: Yale University Press.

Prince, Hugh. 1988. "Art and Agrarian Change, 1710-1815." In *The Iconography of Landscape: Essays on the Symbolic Representation, Design and Use of Past Environments*, edited by Denis Cosgrove and Stephen Daniels, 98-118. Cambridge: Cambridge University Press.

Quartich-Wales, H. G. 1931. *Siamese State Ceremonies: Their History and Function*. London: Bernard Quartich.

R, Jenkyns. 1998. *Virgil's Experience: Nature and History: Times, names and places*. Oxford: Clarendon Press, Oxford University Press

Rae, Douglas. 2003. *City: Urbanism and its End*. New Haven: Yale University Press.

Rapoport, Amos. 1969. *House form and culture*. Englewood Cliffs, N.J: Prentice-Hall.

———. 1977. *Human aspects of urban form : towards a man-environment approach to urban form and design / Amos Rapoport*. 1st ed. Oxford New York: Pergamon Press.

Raymond, Andrz. 2000. *Cairo*. USA: Harvard University Press.

Reed, Christopher. 2004. *Bloomsbury Rooms: Modernism, Subculture and Domesticity*. New Haven: Yale University Press.

Reid, Anthony. 1980. "The Structure of Cities in Southeast Asia, 15th-17th Centuries." *Journal of Southeast Asian Studies* no. 11:235-250.

———. 1993. *Southeast Asia in the Age of Commerce*. 2 vols. Ithaca.

———. 1993. *Southeast Asia in the Early Modern Era*. Ithaca: Cornell University Press.

———. 2000. *Charting the Shape of Early Modern Southeast Asia*. Singapore: Institute of Southeast Asian Studies.

Reynolds, Craig J. 1976. "Buddhist Cosmography in Thai History, with Special Reference to Nineteenth-Century Culture Change." *Journal of Asian Studies* no. 35:203-20.

Reynolds, Frank, and Mani B. Reynolds. 1982. *Three Worlds According to King Ruang: A Thai Buddhist Cosmology*. Berkeley: University of California

Rice, Charles. 2001. "Images at the Built Edge." *Architectural Design* no. 71 (3):25-29.

Rosenau, Helen. 1959. *The ideal city in its architectural evolution.* London Routledge.

Ross, Robert, and Gerard J Telkamp. 1985. *Colonial Cities: Essays on Urbanism in a Colonial Context*. Dordrecht, the Netherlands: Martinus Nijhoff.

Rossi, A. 1982. *The Architecture of the City*. USA: MIT.

Rouvillois, Frederic. 2000. "Utopia and Totalitarianism." In *Utopia: The Search for the Ideal Society in the Western World*, edited by Roland Scaher, Gregory Claeys and Lyman Tower Sargeant, 316-32. Oxford, UK: Oxford University Press.

Rowley, George. 1959. *Principles of Chinese Painting*. Princeton: Princeton University Press.

Rykwert, Joseph. 1976. *The Idea of a Town*. New Jersey: Princeton.

———. 1988. *The Idea of a Town*. USA: MIT.

Sadler, Simon 1998. *The situationist city*. Cambridge Mass. : MIT Press

Sadler, Simon. 2005. "Archigram [electronic resource]: architecture without architecture " In. Cambridge, Mass.: MIT Press,. http://library.adelaide.edu.au/item/1681464.

Saggs, H W F. 1965. *Everyday Life in Babylonia & Assyria*. London: Jarrold & Sons Ltd.

Said, Edward W. 1979. *Orientalism*. New York: Vintage.

Sandercock, Leonie. 1998. *Towards Cosmopolis: Plannning for Multicultural Cities*. Chichester: Wiley.

———. 2003. *Cosmopolis II: Mongrel Cities of the 21st Century*. London: Continuum.

SarDesai, D. R. 1994. *Southeast Asian Past & Present*. Boulder and San Francisco: Westview Press.

Sarin, M. 1982. *Urban Planning in the Third World: The Chandigarh Experience*. London: Mansell.

Sassen, Saskia. 1991. *The Global City: New York, London, Tokyo*. Princeton, NJ: Princeton University Press.

———. 1998. *Globalisation and its discontents*. New York: New Press.

———. 2002. *Global Networks, Linked Cities*. New York: Routledge.

Scafi, Alessandro. 1999. "Mapping Eden: Cartographies of the Earthly Paradise." In *Mappings*, edited by Denis Cosgrove, 50-70. London: Reaktion Books Pty Ltd.

Scheurleer, Pauline Lunsingh. 1996. "Rich Remains from Social Anthropological Work in Eighteenth-Century India." *Journal of the History of Collections* no. 8 (1):71-91.

Schon, Donald A., Bish Sanyal, and William J. Mitchell. 1999. *High Technology and Low-Income Communities*. Massachussetts: MIT Press.

Schulz, Juergen. 1978. "Jacopo de' Barbari's View of Venice: Map Making, City Views, Moralised Geography Before the Year 1500." *The Art Bulletin* no. 60 (3):425-474.

Schuyler, Cammann. 1961. "The Magic Square of Three in Old Chinese Philosophy and Religion." *History of Religions* no. 1 (Summer 1961):37-79.

Schwartzberg, Joseph E. 1992. "Introduction to South Asian Cartography." In *The History of Cartography: Cartography in the Traditional Islamic and South Asian Societies.*, edited by J. B. Harley and David Woodward, 295-331. Chicago: University of Chicago Press Pty Ltd.

———. 1994a. "Maps of Greater Tibet." In *Cartography in the Traditional East and Southeast Asian Societies*, edited by J B Harley and David Woodward, 607-688. Chicago: University of Chicago.

———.1994b. "Introduction to Southeast Asian Cartography." In *The History of Cartography: Cartography in the Traditional East and Southeast Asian Societies*, edited by J. B Harley and David Woodward, 689-700. Chicago: The University of Chicago Press.

———. 1994c. "Cosmography in Southeast Asia." In *The History of Cartography: Cartography in the Traditional East and Southeast Asian*

Societies, edited by J B Harley and David Woodward, 701-740. Chicago: University of Chicago.

———. 1994d. "Southeast Asian Geographical Maps." In *Cartography in the Traditional East and Southeast Asian Societies*, edited by J B Harley and David Woodward, 741-827. Chicago: University of Chicago.

———. 1994e. "Conclusion to Southeast Asian Cartography." In *History of Cartography: Cartography in the Traditional East and Southeast Asian Societies*, edited by J B Harley and David Woodward, 839-842. Chicago USA: The University of Chicago Press.

Scott, Allen John. 2001. *Global City Trends: Trends, Theory, Policy*. Oxford: Oxford University Press.

Scott, James C. 1998. *Seeing Like a State: Why Certain Schemes to Improve the Human Condition Have Failed (The Institution for Social and Policy Studies at Yale University)* New Haven: Yale University Press.

Scribner, R. 1981. *For the sake of the simple folk. Popular propaganda for the German Reformation*. Cambridge: Cambridge University Press.

Seng, Eunice. 2011. "Politics of Greening: Spatial Constructions of the Public in Singapore." In *Non West Modernist Past*, edited by S. W. Lim. William and Jiat-Hwee Chang, 143-160. Singapore: World Scientific Publishing Company.

Seymour, I. 1989. "The Political Magic of John Dee." *History Today* no. 39:29-35.

Shane, Grahame. 2006. "The Emergence of Landscape Urbanism." In *The Landscape Urbanism Reader*, edited by Charles Waldheim, 055-067. New York: Princeton Architectural Press.

Short, John Rennie. 2003. *The World Through Maps: A History of Cartography*. Toronto, Ontario, Canada: Firefly Books.

Shorto, H. L. 1978. "The Planets, the Days of the Week and the Points of the Compass: Orientation Symbolism in 'Burma'." In *Natural Symbols in South East Asia*, edited by G. B. Milner. London: School of Oriental and African Studies.

Shu-hua, Li. 1956. "The South-pointing Carriage and the Mariner's Compass." *Tsing Hua Journal of Chinese Studies* no. 1 (1).

Sigler, Jennifer 1995. *Small, medium, large, extra-large / Office for Metropolitan Architecture Rem Koolhaas and Bruce Mau* New York: Monaccelli Press.

Simpson, J A, and E S C Weiner. 1989. *The Compact Oxford English Dictionary.* second ed. Oxford: Oxford University Press.

———. 1989. The Oxford English Dictionary. In *The Compact Oxford English Dictionary*, edited by T J Benbow. Oxford: Clarendon Press.

———. 1989. The Oxford English Dictionary In *The Compact Oxford English Dictionary*, edited by Editorial Board Members. Oxford: Clarendon Press.

Sinisgalli, Rocco. 2011. *Leon Battista Alberti : On painting : a new translation and critical edition*. Cambridge, New York: Cambridge University Press.

Sircar, D. C. 1971. *Studies in the Geography of Ancient and Medieval India* India: Motilal Banarsidass.

Sitte, Camillo. 1945. *The art of building cities; city building according to its artistic fundamentals*. Translated by Charles T. Stewart. New York: Reinhold Pub. Corp.

Siulak, Sivarska. 1985. *Siamese Resurgence. Asian Cultural Forum Development.* Bangkok: Sukhsit Siam.

Smith, Donald Kimball. 2008. *The Cartographic Imagination in Early Modern England.* Aldershot, Hampshire, England: Ashgate.

Smith, Michael Peter. 2001. "Power in Place: Retheorising the Local and the Global in Urban Studies." *Key note address*.

———. 2001. *Transnational Urbanism: Locating Globalisation*. Oxford, UK: Blackwell.

Smith, Richard J. 1991. *Fortune-Tellers and Philosophers: Divination In Traditional Chinese Society (Westview Special Studies on China)*: Westview Press.

Smithson, Alison Margaret, and Peter Smithson. 2004. *The Charged Void: Urbanism*. New York: Monacelli Press.

Snyder, John P. 1993. *Flattening the Earth: Two Thousand Years of Map Projections*. Chicago: University of Chicago Press.

Sobel, Dava, and H. Andrewes. 1998. *The Illustrated Longitude* New York: Walker.

Soderstrom, Ola. 1996. "Paper Cities: Visual Thinking in Urban Planning." *Ecumene* no. 3:249-81.

Soekmono, R. 1988. *Introduction to the History of Culture Indonesia* 2nd ed. New York: Publishers Canisius. Original edition, 1973. Reprint, 5th.

Soja, Edward W. 2000. *Postmetropolis: Critical Studies of Cities and Regions*. Oxford: Blackwell Publishers.

Solnit, Rebecca. 2001. *Wanderlust: A History of Walking*. London: Verso.

Soloman, David. 2003. *Global City*: Island Press.

Solomon, Daniel 2003. *Global city blues* Washington [D.C]: Island Press.

Soper, Alexander C. 1941. "Early Chinese Landscape Painting." *Art Bulletin* no. 23:141-64.

Spivak, Gayatri Chakravorty. 1999. *A Critique of Postcolonial Reason: Toward a History of the Vanishing Present*. USA: Presidents and Fellows of Harvard College.

Stafford, Barbara Maria. 1984. *Voyage into Substance: Art, Science and the Illustrated Travel Account, 1760-1840*. Cambridge, MA: MIT Press.

Starling, Roger. 1998. "Rethinking the Power of Maps: Some Reflections on Paper Landscapes." *Ecumene* no. 5:105-8.

Steinhardt, Shatzman Nancy. 1998. "Mapping the Chinese City: The Image and the Reality." In *Envisioning the City - Six Studies in Urban Cartography*, edited by David Buisseret, 181. Chicago: The University of Chicago Press Ltd.

Suarez, Thomas. 1999. *Early Mapping of Southeast Asia*. Singapore: Periplus Editions (HK) Ltd.

Sullivan, Michael. 1962. *Birth of Landscape Painting in China: Volume 1 of California studies in the History of Art*. Vol. 1: Routledge Kegan & Paul.

———. 1974. *The Three Perfections: Chinese Painting, Poetry and Calligraphy*. London: Thames and Hudson.

———. 1984. *The Arts of China*. 3 ed. Berkeley and Los Angeles: University of California Press.

Swyngedouw, E. 1999. "Modernity and Hybridity: Nature, Regeneracionismo, and the Production of the Spanish Waterscape, 1890-1930." *Annals of the Association of American Geographers* no. 89 (3):443-465.

———. 2004. *Glocalisations*. Philadelphia, PA: Temple University Press

Sykes, A. Krista 2010. *Constructing a new agenda [electronic resource] : architectural theory 1993-2009* New York: Princeton Architectural Press.

Tafuri, Manfredo. 1976. *Architecture and Utopia*. Cambridge MA: MIT Press.

Tagliacozzo, Eric. 2005. "Tropical Spaces, Frozen Frontiers: The Evolution of Border Enforcement in Nineteenth-Century Insular Southeast Asia." In *Locating Southeast Asia: Geographies of Knowledge and Politics of Space*, edited by Paul H. Kratoska, Raben, Remco. , Nordholt, Henk Schulte, 149-174. Singapore: Singapore University Press.

———. 2007. "An Urban Ocean: Note on the Historical Evolution of Coastal Cities in Greater Southeast Asia." *Journal of Urban History* no. 33 (6):911-932.

Tan, Liok Ee. 2002. Conjunctures, Confluences, Contestations: A Perspective on Penang History. Paper read at The Penang Story-International Conference 2002, 18-21 April 2002, at Penang Malaysia.

Taylor, Keith Weller. 1983. *The Birth of Vietnam*. Berkeley and Los Angeles: University of California Press.

Temple, Robert R. 1998. *The Genius of China: 3000 Years of Science, Discovery and Invention*: Prion.

Thomsen, Christian W. 1994. *Visionary architecture : from Babylon to virtual reality* Translated by John William Gabriel. Munich ; New York Prestel.

Thrower, Norman J. 1996. "Maps and Civilisation: Cartography in Culture and Society." In. Chicago: University of Chicago Press.

Tibbetts, Gerald R. 1979. *A Study of the Arabic Texts Containing Material on South-East Asia (for the Royal Asiatic Society)* London: E. J Brill.

Tickell, Alex. 2004. "Negotiating the Landscape: Travel, Transaction and the Mapping of Colonial India." *The Yearbook of English Studies* no. 34 (Nineteenth Century Travel Writing):18-30.

Tin, U Maung Maung, and Thomas Owen Morris. 1966. "Mindon Min's Development Plan for the Mandalay Area." *Journal of the Burma Research Society* no. 49 (1):29-34.

Tobler, W. R. 1966. "Medieval Distortions: The Projections of Ancient Maps." *Annals of the Association of American Geographers* no. 56:351-60.

Tuan, Yi-Fu. 1974. *Topophilia: A Study of Environmental Perception, Attitudes, and Values*. New Jersey, USA: Prentice-Hall Inc.

———. 1976. "Geopiety: A Theme in Man's Attachment to Nature and Place." In *Geographies of the Mind: Essays in Historical Geosophy*, edited by David

Lowenthal and Martyn J Bowden, 11-39. New York: Oxford University Press.

Turner, Tom. 1996. *City as Landscape: A Post-Postmodern View of Design and Planning*. London: E & FN Spon.

Tyacke, S. 1983. *English Map-Making 1500-1650*. London: British Library Publishing Division.

Ucko, Peter, Ruth Tringham, and G W Dimbleby. 1972. *Man, Settlement, Urbanism*. London: Duckworth.

Urry, John. 1995. *Consuming Places*. London: Routledge.

Vickers, B. 1984. *Occult and Scientific Mentalities in the Renaissance*. Cambridge: Cambridge University Press.

W, Armstrong, and Terry McGee. 1980. "A Theory of Urban Involution." In *Sociology of Southeast Asia: Readings on Social Change and Development*, edited by Hans-Dieter Evers. Kuala Lumpur: Oxford University Press.

Wadley, Reed. 2005. *Histories of the Borneo Environment*. Leiden.

Waldheim, Charles. 2006. *The Landscape Urbanism Reader*. New York: Princeton Architectural Press.

———. 2010. "On Landscape, Ecology and Other Modifiers to Urbanism." *Topos: European Landscape Magazine* no. 71 (Landscape Urbanism):21-24.

Wales, Horace Geoffrey Quaritch 1959. "The Cosmological Aspect of Indonesian Religion. " *Journal of the Royal Asiatic Society of Great. Britain and Ireland* no. 91:100-39.

Wallace, Andrew Russel. 1962 (1869). *The Malay Archipelago*. New York: Dover Publications.

Wang, Ling. 1959. *Mathematics and the Sciences of the Heavens and the Earth*. Vol. 3. Cambridge: Cambridge University Press.

Ward, Graham, and Michel de Certeau. 2000. *The Certeau Reader*. Oxford: Blackwell Publishers.

Watson Andaya, Barbara. 1993. *To Live as Brothers: Southeast Sumatra in the Seventeenth and Eighteenth Centuries*. Honolulu.

Webster, C. 1982. *From Paracelsus to Newton: Magic and the Making of Modern Science*. Cambridge: Cambridge University Press.

Weller, Richard. 2008. "Planning by Design: Landscape architectural scenarios for a rapidly growing city." *Journal of Landscape Architecture* no. Autumn 2008:6-17.

———. 2009. *Boom Town 2050: Scenarios for a Rapidly Growing City*. Perth, Australia: University of Western Australia Publishing.

Wells, J K , and John Villiers. 1990. *The Southeast Asian Port and Polity: Rise and Demise*. Singapore.

Werbner, Pnina. 1999. "Global Pathways. Working Class Cosmopolitans and the Creation of Transnational Ethnic Worlds." *Social Anthropology* no. 7 (1):17-35.

Wertheim, W F. 1980. "Changing Southeast Asian Societies: An Overview." In *Sociology of Southeast Asia: Readings on Social Change and Development* edited by Hans-Dieter Evers, 8-24. Kuala Lumpur: Oxford University Press.

Wheatley, Paul. 1954. "A Curious Feature in Early Maps of Malaya." *Imago Mundi* no. 11:67-72.

———. 1961. *The Golden Cheronese: Studies on Historical Geography of the Malay Peninsula Before AD1500*. Kuala Lumpur.

———. 1972. *The Pivot of the Four Corners: A Preliminary Enquiry into the Origins and Character of the Ancient Chinese City*. Edinburgh: Edinburgh University Press.

———. 1983. *Nagara and commandery: Origins of the Southeast Asian Urban Traditions*. Chicago, Illinois, USA: The University of Chicago.

Whitfield, Peter. 2010. "The Image of the World: 20 Centuries of World Maps." In. London: The British Library.

Whitmore, John K. 1994. "Cartography in Vietnam." In *The History of Cartography: Cartography in the Traditional East and Southeast Asian Societies*, edited by J. B Harley and David Woodward, 478-508. Chicago: The University of Chicago Press.

Widodo, Johannes. 2004. *The Boat in the City: Chinese Diaspora and the Architecture of Southeast Asian Cities*. Singapore: Marshall Cavendish Academic.

Wigley, Mark. 1998. *Constant's New Babylon: The Hyper-Architecture of Desire*. Rotterdam: 010 Publishers.

Wilson, Constance. 1982. "Cultural Values and Record Keeping in Thailand." *CORMOSEA [Committee on Research Materials on Southeast Asia] Bulletin* no. 10 (2):2-17.

Winichakul, Thongchai. 1994. *Siam Mapped: A History of the Geo-Body of a Nation*. USA: University of Hawai'i Press.

Withers, Charles W. J. 1999a. "Geography, Enlightenment, and the Paradise Question." In *Geography and Enlightenment*, edited by David N. Livingstone and Charles W. J. Withers, 67-92. Chicago: University of Chicago.

Withers, Charles W J, and David N Livingstone. 1999b. "Introduction: on Geography and Enlightenment." In *Geography and Enlightenment*, edited by Charles W J Withers and David N Livingstone, 1-28. Chicago, USA: The University of Chicago Press.

Wolff, Janet. 1981. *The Social Production of Art*. New York: New York University Press.

Wolters, O W. 1967. *Early Indonesian Commerce: A Study of the Origins of Srivijaya*. Ithaca.

Wood, Denis. 1992. *The Power of Maps*. New York: Guilford Press.

Woodward, David. 1985. "Reality, Symbolism, Time, and Space in Medieval World Maps." *Annals of the Association of American Geographers* no. 75 (4 Dec 1985):510-521.

———. 1994. "Preface." In *The History of Cartography: Cartography in the Traditional East and Southeast Asia.*, edited by J. B Harley and David Woodward, xxiii-xxvii. Chicago, USA.: The University of Chicago Press Pty. Ltd.

Wright, Gwendolyn. 1991. *The Politics of Design in French Colonial Urbanism*. Chicago USA: The University of Chicago Press.

Wright, John K. 1947. "Terrae Incognitae: the place of imagination in geography." *Annals of the Association of American Geographers* no. 37:1-15.

Wu, Liangyong. 1986. "A Brief History of Ancient Chinese City Planning." *Urbs et Regio* no. 38.

Xiaocong, Li. 2004. A Descriptive Catalogue of the Traditional Chinese Maps Collected in the Library of Congress. edited by Library of Congress: The University of Virginia.

Yamashita, Shinji, and J. S. Eades. 2003. *Globalisation in Southeast Asia: Local, National and Transnational Perspectives*. New York USA: Berghahn Books.

Yeang, Ken. 1997. The Skyscraper Bioclimatically Considered. In *Lecture Series accompanied by an exhibition of T. R Hamzah and Yeang's Architecture*. New York Chapter: American Institute of Architects.

Yee, Cordell D K. 1996. *Space and Place: Map-making East and West : Four Hundred Years of Western and Chinese Cartography*: Saint John's College Press.

Yee, D K, Cordell. 1994. "Chinese Cartography among the Arts: Objectivity, Subjectivity, Representation." In *History of Cartography: Cartography in the Traditional East and Southeast Asian Societies*, edited by J. B. Harley and David Woodward, 128-169. Chicago: University of Chicago.

Yeo, Richard. 1991. "Reading Encyclopedias: Science and the Organisation of Knowledge in British Dictionaries of Arts and Sciences, 1730-1850." *Isis* no. 82:24-49.

Yeoh, Brenda S. A. 1996. *Contesting Space: Power Relations and the Urban Built Environment in Colonial Singapore*. New York: Oxford University Press.

Yeung, Y M, and C P Lo. 1976. *Changing Southeast Asian Cities: Readings on Urbanisation*. Oxford.

Yu, Kongjian. 2010. "Five Traditions for Landscape Urbanism Thinking." *Topos: European Landscape Magazine* no. 71 (Landscape Urbanism):58-63.

Zukin, S W. 1991. *Landscapes of Power. From Detroit to Disney World, Berkeley, Los Angeles*. Oxford: University of California Press.

INDEX